TEACHING THE ESSENTIALS OF ARITHMETIC

by

Philip Boswood Ballard

YESTERDAY'S CLASSICS

ITHACA, NEW YORK

Cover and arrangement © 2022 Yesterday's Classics, LLC.

This edition, first published in 2022 by Yesterday's Classics, an imprint of Yesterday's Classics, LLC, is an unabridged republication of the text originally published by University of London Press Ltd. in 1928. For the complete listing of the books that are published by Yesterday's Classics, please visit www.yesterdaysclassics.com. Yesterday's Classics is the publishing arm of Gateway to the Classics which presents the complete text of hundreds of classic books for children at www.gatewaytotheclassics.com.

ISBN: 978-1-63334-177-7

Yesterday's Classics, LLC
PO Box 339
Ithaca, NY 14851

TO

The Young Arithmetician

WHOM

I KNOW BEST AND LOVE BEST—

MY DAUGHTER

BRONWEN

A few strong instincts and a few plain rules.

WORDSWORTH.

Pursued in the spirit of a philosopher and not of a shop keeper, arithmetic has a very great and elevating effect, compelling the soul to reason about abstract number, and rebelling against the introduction of visible or tangible objects into the argument.

PLATO: *The Republic.*

And what of all this? Marry, read the book and you shall know; but read nothing except you read all. And why so? Because the beginning shews not the middle, and the middle shews not the latter end.

THOMAS DELONEY: *The Gentle Craft* (1543-1600).

PREFACE

RARELY is a teacher satisfied with the arithmetic of his class. He is not allowed to be. Somebody always has something to say about it—generally something unpleasant. If it is not a colleague, it is probably an inspector; if it is neither of these, it is a parent or an employer. And even when the criticism is not aimed at a particular person, but lies couched in a general report on a public examination, or in a survey of a prescribed area, or in an utterance from the platform or the press, its generality makes it none the less a disturbing force. The circle of disturbance is larger, that is all. Like slugs fired from a blunderbuss, the discharge hits many though it kills none. Not that one can object to these general criticisms. On the whole they do good. What the teacher has to do is to grow a skin thin enough to let him know when he is hit, but thick enough to protect him from serious wounds. It is true that many of these blunderbuss criticisms scatter through the empty air and hit nothing (this is specially true of those that come from the employer and the press), but some of them reach home. Some of them touch real points of weakness. And these, as a rule, are the charges that are common to all the indictments.

What the critics say, and say with one voice, is that the fundamentals are weak; that our children are badly grounded in the very ABC of arithmetic. They don't complain that the pupils can't work compound proportion; but they do complain that they can't work compound multiplication. Nobody grumbles at their not knowing logarithms; but everybody grumbles at their not knowing the multiplication table, and not being able to add and subtract with that ease and accuracy which ordinary life demands.

There is no lack of excellent textbooks on the market. There are Common-sense Arithmetics which are chock full of common sense, and Efficiency Arithmetics which cannot fail to make a lad efficient if he can only be induced to work the examples. And there are many others equally good, and, to all appearances, equally worthy of their titles. Nor is there a lack of good books on the teaching of arithmetic. It is difficult to conceive a clearer or more comprehensive book than Mr. F. F. Potter's, or a more helpful and charming little book than Miss Jeannie B. Thomson's. And yet our pupils bungle at the very rudiments of the subject. The only conclusion I can come to is that the textbooks are too good for the pupils. The arithmetical fare we offer them is ill-suited to their digestions: it is either too rich or too much. What they need is a simpler diet, a diet which they can manifestly turn into sound bone and muscle, a diet for which a keen appetite is the best sauce. From what I can discover, the appetite at present is none too good.

As for our books on method, they all suffer from an excess of logic and a dearth of psychology. They follow tradition in stressing the rational side of arithmetic. After making a few concessions to the immature mind, which may be allowed to gain its first notions of number from bricks or beans, they hasten to apply general principles to particular instances. They proceed deductively. Every step at every stage has to be reasoned out. No unexplained process is allowed to be taken on trust and used on the sole ground that the process works. No lumps of knowledge, however useful they may be as they are, are allowed, even for a while, to escape the grinding of the logical machine. And arithmetic is almost made to appear as a mere branch of deductive logic.

This is the English view. But there is another view which finds its clearest utterance in Thorndike's book on *The Psychology of Arithmetic*. This book sets the tune to which all American educators now dance. Arithmetic is presented as an inductive science. Reasoning starts with the concrete fact and ends with the concrete fact. Children learn arithmetic by working sums. The justification for the mode of procedure is that the answer is right. The ground for believing the answer to be right is the word of the teacher, or the result got by reversing the process, or, in the last resort, the irrefutable evidence afforded by counting. The child does a thing first and understands it after. Doing is the important thing; and practice in doing—the practice that, "line upon line, here a little and there a little," fixes deeper and deeper a series of habits. Arithmetic is in fact not so much an application of broad general principles as as organisation of habits.

Thus we have on the one hand the English view that arithmetic is logic, and the American view that it is habit. The contrast is interesting and significant; but it is not new. It resembles, in fact, the antithesis between the Platonic and the Aristotelian views of virtue. To Plato virtue is knowledge; to Aristotle it is habit. To Plato it is an intellectual grasp of the consequences of our acts; to Aristotle it is the practice of choosing the mean between two extremes.

These differences in emphasis and in outlook are not of mere theoretical import: they vitally affect practice. They prescribe what we shall teach, how we shall teach it, and how we shall test it. Hence the study of these two contrasting attitudes may yield us the key to the solution of our difficulties. I think the key is really there; but it is a duplicate key, or rather a multiplicate key. I had myself on certain vital points come to pretty much the same conclusions as Thorndike long before I knew what Thorndike's conclusions were. So no doubt had many of my readers. The outstanding fact is that the habit element in arithmetic has in recent years in England been obscured by ill-founded views on the place and function of intelligence in the study of the subject. And one of the main aims of this little book is to remove the obscurity, and to reveal the role of habit in the mastery of arithmetic; to show that in the erection of a sound fabric of knowledge, though intelligence may be the architect and builder, yet it is habit that gives cohesion to the bricks and adhesion to the mortar, and to attempt to build without its aid is worse than trying to build *on* sand: it is trying to build *with* sand.

Although this book aims at reconciliation, it must not be thought that I am mainly concerned in balancing pros and cons, or in expanding the sentence: "Much may be said on both sides." I am not. I don't sit on the fence: I come down definitely on one side or the other. Let me cite a few instances. I think the method of teaching subtraction by decomposition a vicious method. I am convinced that the policy of shirking long division till late in the course and substituting division by factors is wasteful and ineffective. I hold that compound multiplication by factors is a clumsier method than direct multiplication in one line, and that the unitary method of working proportion is more cumbersome that the fractional method. And I am a great believer in the King's highway—in having one good standard method of working a given type of sum. I have observed that those who show too eager a desire to avoid the beaten track and discover short cuts often come to grief. They either lose their way or arrive late. Meanwhile their more pedestrian classmates who have gone the longest way round have really found the nearest way home.

We must distinguish (if we can) between new ideas that have come to stay and new ideas that arise from chance and change—ideas which are in the fashion, and have in consequence a certain air of smartness, but come off badly when subjected to the wear and tear of the classroom. The worst of it is, when one idea gets into fashion it pushes another idea out of fashion. And the other is often the better of the two. It was the fate that befell "equal additions" when "decomposition" got into vogue; it is a disaster that threatens to befall that good old method of multiplying decimals—counting the decimal places., The ground of the threatened taboo seems to be that if this old method is continued, children will be getting their sums right all over the place, and there will be no need for them to acquire some particular piece of pedantic ritual. There are other rules, too, which have recently come under the ban of the doctrinaire, rules such as proportion, practice, alligation, and the reduction of problems to types. Such bans are, as a rule, quite unreasonable: they have no basis in theory, no justification in practice.

This bold (and bald) confession of faith may seem to contain a touch of defiance: I may seem to be trailing my coat. In reality I am merely trying to be honest, trying to tell the reader what he may expect to find in the pages that follow. The one question which I have steadily kept before me is: What are the methods which, while theoretically sound, succeed best with children? What, in other words, are the methods which enable the young learner to get the most sums right in the shortest time? The evidence I seek is that of experiment and experience, and my court of appeal is the classroom.

There is no branch of study which is so dominated by examinations as arithmetic. For it is the most examinable subject in the curriculum; and being the most examinable it is the most examined. Whatever the type of general examination, arithmetic is sure to be brought in. Even the intelligence examination is not exempt. Indeed, it has introduced a new model—a Parisian model, popularised if not invented by Alfred

Binet. The technique of the mental test demands that a question should be brief, simple, and unequivocably scorable. It should, if possible, carry one mark or none. This technique has already begun to influence the examination question in arithmetic: it has reduced its size and its complexity. And in so doing it has led to the invasion of new territory. Let us look for a moment at the old order of things; which is still, in the main, the prevailing order of things. The traditional test in arithmetic consists of a number of questions, each of which takes from ten to fifteen minutes to answer. The examination sum is a ten-minutes sum. If another paper is set, it consists of mental arithmetic questions with about ten seconds allowed for each. The range between the ten-seconds sum and the ten-minutes sum is entirely untouched. Here is a huge tract of No-man's-land the existence of which is ignored by the examiner. And yet in everyday life this is the very region in which most of our calculations lie. If it be objected that the neglect applies to examination sums only, it may be replied that classroom sums are echoes of examination sums; for nearly all the arithmetic done in our schools (I state it as a fact, not as a fault) is done in preparation for some sort of examination. A change in the examination must mean a change in the teaching, and a book which deals with teaching must take cognisance of the fact.

I do not propose to discuss why we teach arithmetic, because that is a question which nobody really asks with a genuine desire to hear the answer. When he asks it at all, it is with a desire to tell the answer, not to hear it. The teacher never asks it. He knows he has to teach the subject in any case, and he is content to leave it at that. But there are other questions which he is constantly asking. How, for instance, may John Smith, a decent enough lad in his way, but one who seems to regard arithmetic as something "from which the mind instinctively retires"—how may John Smith be brought to take an interest in the subject? The question of motivation in fine is not fictitious: it is real, vital, and of perennial urgency. Pedagogical textbooks ignore it; they assume that John Smith does not exist; they teach in effect that little children take to arithmetic as ducks take to water. All we have to do is to deal out to the little dears a nice set of sums, and they will work them with avidity, and even ask for more. And their minds are all agog for explanations, eager to know the why and the wherefore of all the processes they employ. No teacher holds these views. Many of them indeed are filled with surprise when they find a child who will cipher, not as a task, but as a joy, and they will secretly point him out as a prodigy—as something strange and unnatural. And yet the subject is full of romance, as Professor Spearman has recently been pointing out. "How comes it," he asks, "that this mathematics, controller of destinies, source of delight, fount of emotion, breeder of romance, has arrived at being almost universally besmirched with the attribute of 'dull'?" He goes on to say: "For my part, I would mainly attribute this huge miscarriage of justice to the manner in which the subject is being taught in school." (*The Outlook*, December 24, 1927). That Professor Spearman is right I have little doubt. But whether he is right or wrong, he raises a question which clamours for an answer.

As I have already said, I have attempted a reconciliation—or, if you Jike, a compromise—between the extreme English view and the extreme American view. There is another *entente* which I have at heart: a more complete understanding between the primary school, and the secondary school, and a better co-ordination of methods. The methods are sometimes sadly at variance. In the primary school, for instance, the pupil is taught to begin to multiply with the units figure of the multiplier; in the secondary school he is taught to begin at the other end. This seems a small matter, but it is big with consequence to the higher half of the arithmetic course. And in this instance the experimental evidence is on the side of the secondary school. There is a distinct gain in beginning to multiply from the weightier side.

At other times it is the primary school that is right and the secondary school wrong—right and wrong being determined as before by experimental evidence. Primary-school children are taught a simple and extremely efficient method of multiplying decimals— the method of adding decimal places. It is a sound, rational method, easy to teach, easy to understand, easy to apply. It has all the desirable characteristics of a standard rule-of-thumb method. It depends, too, upon a principle which is of wide application in the higher branches of arithmetic. And yet, for some reason or other, this sane and simple method is in many secondary schools regarded as arbitrary and irrational, and is as rigidly suppressed as though it were some deadly heresy destructive of all mathematical integrity of mind. And in its place is put a cumbersome method of multiplying known as the "standard form" methbd—a method which takes nearly twice as long and is about half as accurate as the old-fashioned method. At first I thought there was in "standard form" some mysterious virtue which had escaped my notice. But after discussing the matter with the best mathematicians I know, I find there isn't. I was assured by one master that it was a method invented to prevent parents from helping children with their homework. The real purpose, however, was to secure uniformity of method among entrants to secondary and public schools. The inventors of the method should try again, and invent a better one.

The obligations I have to acknowledge are many. The writers to whom I am chiefly indebted are Augustus De Morgan (the delight of the connoisseur), Sir Oliver Lodge, and Professors Spearman, Nunn, and Thorndike. The main stimulus, to the writing of this book, however, has been my own experience in schools. Such views as I put forward are the harvest of many years of teaching, testing, observing, and experimenting. The fact that I have more respect for experiment than for opinion—including, I hope, my own opinion—does not prevent me from being inordinately pleased when I find myself in agreement with my friends; especially when those friends happen to be both mathematicians and philosophers. I am happy therefore to record that Professor Spearman has read in manuscript the chapters on mathematical reasoning, and Professor Nunn the chapter on incommensurables, and that both concur generally in the views therein expressed.

I am deeply indebted to my colleague Mr. E. P.

Bennett, who has read through the manuscript with much care, and has given me the benefit of his knowledge and experience. He has also been kind enough to correct the proofs. It is impossible to express adequately my appreciation of the ready and generous help I have received in my arithmetic investigations for this and other books from teachers of all grades. Chief among them are Mr. A. Wisdom, Mr. J. G. Robson, Mr. T. H. Elliott, Mr. H. R. Neilson, and Mr. H. H. Spratt. Finally, I owe an accumulated debt of gratitude to Mr. W. Stanley Murrell, the Manager of the University of London Press, who has wisely counselled me for many years, and has with great skill piloted a number of my books through the straits of printing and binding out into the open sea.

P. B. BALLARD.

CHISWICK.

April 1928.

CONTENTS

CHAPTER I

INTELLIGENCE AND HABIT

What a piece of work is man! How noble in reason!
how infinite in faculties!
Shakespeare: Hamlet.

A house built upon reason is a house built upon sand.
Knowledge must become automatic before we are safe
with it.
Samuel Butler: Life and Habit.

Whenever thought is necessary, it is to be exercised
vigorously, but it should not be wasted over simple mechanical
operations.
Sir Oliver Lodge: Easy Mathematics.

A generation ago many of the most vocal members
of our profession made much ado about Intelligence. It
was their great word. They used it in all their arguments;
they stressed it in all their speeches. The cultivation
of intelligence was the grand aim and purpose of
education, and the value of each branch of study was
to be measured by the extent to which it furthered this
great purpose. And a method of teaching was good or
bad according as it hastened or hindered the growth of
intelligence. Intelligence had thus become a touchstone

as well as a watchword. It is true that the meaning of the word was a little obscure, and the sense in which it was used was prone to vacillate. Sometimes it meant this, and sometimes it meant that. In a general way, however, it included the higher, the more rational, the more distinctively human operations of the mind, and excluded those powers which we share with the beasts that perish. And as intelligence was the noblest function of the human mind, so was memory, in all its manifestations, the most ignoble. Doubly ignoble was it when it took the form of habit memory—memory that had become so deeply embedded in brain and nerve as to be organic, to be part and parcel of the human organism. For by this time the process had become so mechanical that it almost worked of its own accord. Man made in God's own image had been reduced to the level of a machine. And so, while intelligence was glorified, automatism was vilified; while one was lauded to the skies, the other became an object of derision and scorn.

It was an inspiring doctrine, well calculated to capture the young and generous mind. The very catchwords and slogans were full of appeal: "Capacity, not content"; "Power which brings knowledge, not knowledge which may or may not bring power"; "An agile mind rather than a full mind"; "The only habit which the child should be allowed to form is the habit of contracting no habit at all" (this from Rousseau's *Émile*). These are attractive sentiments. The only objection to them is that they won't work. They are out of touch with reality. They are based on a false notion of human nature, and of the rôle of intelligence on the one hand and of habit on the other in the building up of human knowledge. The result is that they either do not influence practical teaching at all, or influence it to its hurt.

The doctrine involves the belief that brain-power can be generated by special studies—that mental gymnastics of a particular kind can produce a mental gymnast of a general kind. The technical term for this theory is "formal training." The first man to shake English teachers out of their complacent trust in this theory was Sir John Adams, then plain John Adams of Glasgow. In his little book on *Herbartian Psychology* published in 1897 he subtly and humorously insinuated such doubts into the minds of his readers that they soon ceased to talk about cultivating the faculty of intelligence. Others followed his lead. They took his matter though they changed his manner. Dr. Hayward preached the new Herbartian creed from the platform and from the professional press with all the ardour of an old Hebrew prophet. As Jonah predicted the downfall of Nineveh, so did Hayward proclaim the downfall of Formal Training, and many of his hearers thought he was talking nonsense; they now know he was talking plain common sense—with perhaps a touch of exaggeration to make it picturesque. Meanwhile Mr. Winch, with a keen eye for the essentials of a problem, had been putting the matter to the test of experiment. He re-christened the problem and called it the problem of transfer. The question at issue appeared in this form: Is ability acquired in one function transferred to another function, distinct and dissimilar? A function gains with practice. Does it share the gain with others, or does it keep it all for itself? Mr. Winch's conclusions were that though it keeps the huge bulk for itself, it does probably share a little of it with others. If there is any transfer, it is very small.

In the meantime, researches of a different kind were being pursued by Alfred Binet in France and by Professor Spearman and Dr. Cyril Burt in England. These pioneers were soon joined by other workers in other lands, all intent on one task, the task of measuring that mysterious something which went under the name of intelligence. Absurd as seemed the attempt to measure so elusive a thing, the researchers hoped that the very attempt to measure it would enable them to identify it, to pin it down, to define its nature and its limits. And their hopes have largely been realised. They have been able to show that intelligence manifests itself most unmistakably in those processes which are commonly called reasoning; that it is scarcely influenced at all by environment and training; and that it mainly determines a child's capacity to profit by the lessons he receives at school. Incidentally it has been brought to light that this "intelligence" of the psychologist is not quite what the man in the street means by intelligence.

What then becomes of the theory that the main object of teaching arithmetic, or indeed teaching anything else, is the cultivation of intelligence? When a teacher claims that he is achieving this object he does not quite mean what he says. What he really means is that he makes his pupils use what intelligence they already have—makes the child with five talents use his five, and the child with one talent use his one. That is the most he can do. He cannot by any species of jugglery turn a one-talent child into a two-talent child.

Intelligence is the mind's original capital (as Dr. William Garnett once phrased it), and the vital question is: Does it accumulate simple interest or compound interest? The out-and-out opponents of formal training— whole-hoggers like Dr. Sleight—cling to the simple-interest theory. More cautious investigators incline to the compound-interest theory—with this important reservation: the amount of interest that is added to the original capital is very small in comparison with the capital itself. The whole question is complicated by the fact that intelligence is not the whole of the mind's native endowment. A man is born with certain specific abilities as well as general ability; and these specific abilities are eminently trainable. I am here, however, dealing with general ability only, and am trying to show that the teacher's main concern with the intelligence of his children is to ensure that none of the mind's capital is unused, that none of the talents is wrapped up in a napkin.

As the Victorian theorists' views on the cultivability of intelligence were wrong, so also were their views on the worthlessness of habit. It would not be difficult to turn the tables on the last generation and sing the praises of habit at the expense of intelligence. Indeed, one of their own contemporaries has already done so. Samuel Butler's *Life and Habit,* which appeared in 1878, may be regarded as a long and brilliant essay in praise of automatism. Automatism is the final flower of knowledge—the goal to which all knowledge drives. And knowledge becomes more perfect as it becomes less conscious. The conscious knower is the bungler; the unconscious knower is the expert.

Let us consider what happens when a man learns to do something—to ride a bicycle, for instance. His first attempt is wild and fervid. He begins full of hope, even if he ends full of bruises. His mind is at high tension, his thinking at white heat. And the more he blunders the more furiously he thinks. But in the course of time his failures get fewer and fewer and his success more and more assured. A complex habit is gradually being formed; and as the habit gets stronger, the conscious control gets weaker. As automatism gets driven in, intelligence gets driven out. When the bicycle has been completely mastered it can be ridden with the minimum of attention and the minimum of volition. The whole process has been sinking more and more into the unconscious. That is what the Frenchman had in mind when he described education as the turning of the conscious into the unconscious.

Habit formation is therefore a beneficent thing. Far from being a deadening and enslaving process, it is a process of emancipation. All along the line there is a gradual releasing of energy, which is thus set free to function in new directions. And not only is it liberated, but it is provided, in the habit itself, with new material in which to work. Habit is thus seen to be a means of mental economy. It funds our knowledge and gives it security. So far as effective action is concerned, we are much safer in the hands of habit than in the hands of intelligence. Nature herself has never dared to entrust the vital physiological functions to the control of the intellect. The heart beats and the blood circulates with an automatism which is absolute and complete.

One characteristic of an automatic act is that it goes best by itself. If we poke thoughts into it, it goes wrong. We walk best when we are not thinking about walking; when we think out the order of the various movements by which we dress of a morning, we arrive late at breakfast.

CHAPTER II

EDUCTION

He thrids the labyrinth of the mind.

TENNYSON: *In Memoriam*.

THE reader with a turn for philosophy may justly complain that in the previous pages I have afforded him many glimpses of the familiar and not a few of the obvious. I will now, however, shift the argument to newer ground. As I have already stated, the main business of intelligence is reasoning. But what is reason? The mere fact that we still argue over the question, Can animals reason? indicates the general fogginess of our opinions on the matter. It isn't that we don't know what animals can do, but that we don't know whether to call it reasoning. The one thing upon which all are agreed is that in reasoning we somehow or other derive a new idea from old ideas—and this without further recourse to experience. Whether consciously or unconsciously, whether spontaneously or by deliberate effort, we "draw out" from something that is given something which is not explicitly but only implicitly given. The old term for this "drawing out" is "inference" or "deduction." Professor Spearman has given it the more appropriate term "eduction," and has defined it more closely. And it is this eductive logic of Professor Spearman's that I wish to apply to the reasoning processes in arithmetic. I select it on the simple ground that it is the only sort of logic that fits.

The logic that held the field for over a millennium was the deductive logic of Aristotle. Its essential form is the syllogism, of which the following is simple example:

> All insects have six legs;
> All bees are insects;
> Therefore all bees have six legs.

It was thought that all true reasoning conformed to this type; yet every attempt to press mathematical reasoning into the syllogistic mould has signally failed.

In modern times it has been realised that deductive logic is only part of the process by which we arrive at truth. The larger process is called induction. It is the method of scientific inquiry. It begins earlier than deduction and includes it. It begins, not with the statement that all insects have six legs, but with the evidence for that statement. It begins, in fact, with particulars and not with generalities. The pertinent fact, however, is that it has failed, just as the earlier logic has failed, to explain the essential nature of mathematical reasoning.

The failure of traditional logic is probably due to the fact that it did not carry its analysis far enough: it had not arrived at the elemental units. It had reached the molecules, but not the atoms; or, to be ultra-modern, the atoms but not the electrons. The honour of finding the electrons belongs to Professor Spearman. His electrons are "fundaments" and "relations."

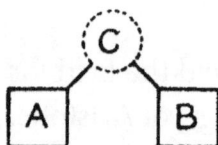

FIGURE 1

Let A in Figure 1 (these diagrams are Dr. Spearman's own) represent the number 2, and B the number 8. Having got a mental grip of A and B, we can cognise or "educe" a relation *(c)* between them. C may be "smaller than," or "¼ of," or "the cube root of," or "6 less than." The actual relation educed will depend on the task the thinker has in hand: he will select the one pertinent to his purpose. If, again, A is 5 + 2 and B is 7, then C is "equal to." This relationship of equality plays an important part in mathematics, The multiplication table is a convenient list of the more useful fundaments bearing this relationship. So are the other tables. In the solution of equations and the simplification of fractions the relationship of equality with its well-known symbol (=) dominates the whole proceedings.

In addition to the first kind of eduction, the eduction of relations, there is a second kind, the eduction of correlates. In Figure 2, A and C being given, we have to educe B, the missing fundament. If, for instance, A is 5 and C is "half of," we know that B must be 10. This is the eduction of a correlate, for the educed fundament

B is a correlate of the given fundament A. Each item of the multiplication table represents an operation which comes under the second principle. In the statement "4 times 7 are 28," 7 is the given fundament, "taken 4 times" is the given relation, and 28 is the educed correlate.

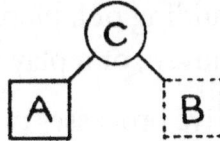

FIGURE 2

It must not be thought that the material on which the two eductive processes work is as simple as the above exposition would lead us to believe; for the product of one eduction may become a fundament of another. If A, B, C, D in Figure 3 are the fundaments from which we start (we must start somewhere), the relations between them may become the basis of new eductions. The figure indicates but a few of the possible relationships; and merely suggests their possible complexity. We may indeed pile up an indefinite number of relations

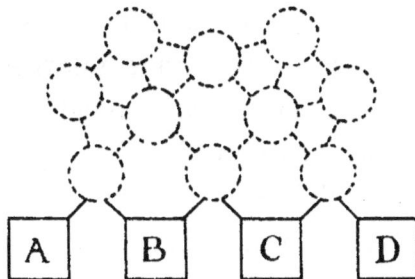

FIGURE 3

and correlates and form a huge fabric of constructive thought which terminates in notions widely removed from the elementary notions from which we started. Nor must it be thought that the construction of this fabric applies to reasoning only. It operates over the whole field of cognition, from the following of a cinema film to the understanding of a picture by Velasquez, a symphony by Beethoven, or a play by Shakespeare.

These two eductive processes pervade the whole of our thinking: they are the two steps by which the mind marches into new realms of thought. Direct experience and memory provide the solid starting-ground, but the forward thrust, the life and movement of the human spirit, come from eduction. As the verb is the soul of the sentence, so is eduction the soul of reason; and as the verb cannot live without the noun, so does eduction owe its very existence to the tributary service of sense and habit. And not habit only—the fixed and facile stage of memory—but every form of memory, every form of reproduction. Our ordinary thinking is a kaleidoscopic medley of eduction and reproduction. It is reproduction that gives the grist; it is eduction that does the grinding. And without grist there can be no grinding. Why is it that we are able to rear in the mind the sort of edifice pictured in Figure 3? Eduction is a narrow process; it cannot occupy a broad stage: it needs the spot-light. And the only way it can build high is by accepting the services—the menial services if you like—of memory. Memory hands over the products of previous eductions and enables the actual eductive activity to work among the higher storeys. If it is kept at work in the basement, the higher storeys will never be reared.

The moral of all which is: Let children learn the multiplication table so that it can be reproduced, item by item, with mechanical precision and promptitude; fix the routine of the simple rules so that they absorb the minimum of creative thought; foster the formation of useful habits so that intelligence may be kept at work in its proper sphere. Habit is a servant; see that it is a good servant. Intelligence is a master; see that it is not allowed to concern itself too much with life below-stairs.

CHAPTER III

DEDUCTION AND INDUCTION

Reasoning is essentially the organisation and control of habits of thought.

Pure arithmetic as it is learned and known is largely an inductive science.

The older pedagogy commanded the pupil to reason and let him suffer the penalty of small profit from the work if he did not. The newer pedagogy secures more reasoning in reality by not pretending to secure so much.

THORNDIKE: *The Psychology of Arithmetic.*

Sure, he that made us with such large discourse,
Looking before and after, gave us not
That capability and god-like reason
To fust in us unused.

SHAKESPEARE: *Hamlet.*

THE last chapter dealt with eduction and the products of eduction. It is the custom in America to call these products bonds. "7 + 8 = 15" is one bond, "7 × 8 = 56" is another. To form a bond is to understand it and to learn it by heart—especially to learn it by heart. Professor Thorndike holds the view that the problem of teaching arithmetic is a problem in "the development of a hierarchy of intellectual habits," and becomes in large measure a question of "the choice of bonds to be formed, and of the best order in which to form them and the best means of forming them in that order." "Bond" is a good word, and useful. It does not allow us to forget that binding has to be done. When the elements are united for the first time in the child's mind—united by the mind's own creative energy—it is an act of eduction. And the very act itself tends to bind the elements together. But the bond is very weak. It has to be strengthened by repetition. And, other things being equal, the more frequent the repetition the stronger the bond. It thus happens that the bonds actually present in the learner's mind are of varying strength; and one of the many problems that confront the teacher is: How strong must a bond be in order that it might be effective as a working habit?

It is clear that the Americans do not underestimate the value of habit, but how do they relate habit to reasoning? We find no such analysis of the very rudiments of reasoning as we find in Spearman, but we do find a rough classification of the methods of arithmetical reasoning into two groups—deductive and inductive. The distinction will become clear by examining the following instances given by Thorndike.

Long Division: Deductive Explanation
To Divide by Long Division

1. Let it be required to divide 34531 by 15.

Operation

Dividend

Divisor 15)34531(2302$\frac{1}{15}$ Quotient
$$\frac{30}{45}$$
$$\frac{45}{31}$$
$$\frac{30}{1 \text{ Remainder}}$$

For convenience we write the divisor at the left and the quotient at the right of the dividend, and begin to divide as in Short Division.

15 is contained in 3 ten-thousands 0 ten-thousands times; therefore, there will be 0 ten-thousands in the quotient. Take 34 thousands; 15 is contained in 34 thousands 2 thousands times; we write the 2 thousands in the quotient. 15 × 2 thousands = 30 thousands, which, subtracted from 34 thousands, leaves 4 thousands = 40 hundreds. Adding the 5 hundreds, we have 45 hundreds.

15 in 45 hundreds 3 hundreds times; we write the 3 hundreds in the quotient. 15 × 3 hundreds = 45 hundreds, which subtracted from 45 hundreds, leaves nothing. Adding the 3 tens, we have 3 tens.

15 in 3 tens 0 tens times; we write 0 tens in the quotient. Adding to the three tens, which equal 30 units, the 1 unit, we have 31 units.

16

15 in 31 units 2 units times; we write the 2 units in the quotient. 15 × 2 units = 30 units, which, subtracted from 31 units, leaves 1 unit as a remainder. Indicating the division of the 1 unit, we annex the fractional expression, $\frac{1}{15}$ unit, to the integral part of the quotient.

Therefore, 34531 divided by 15 is equal to 2302$\frac{1}{15}$.

[B. Greenleaf, *Practical Arithmetic*, '73, p. 49.]

LONG DIVISION: INDUCTIVE EXPLANATION
Dividing by Large Numbers

1. Just before Christmas Frank's father sent 360 oranges to be divided among the children in Frank's class. There are 29 children. How many oranges should each child receive? How many oranges will be left over?

Here is the best way to find out:

$$\begin{array}{r} 12 \\ \overline{29)360} \\ 29 \\ \overline{70} \\ 58 \\ \overline{12} \end{array}$$	*Think how many 29s there are in 36. 1 is right.*
	Write 1 over the 6 of 36. Multiply 29 by 1.
	Write the 29 under the 36. Subtract 29 from 36.
and 12 remainder	*Write the 0 of 360 after the 7.*
	Think how many 29s there are in 70. 2 is right.
	Write 2 over the 0 of 360. Multiply 29 by 2.

Write the 58 under 70. Subtract 58 from 70.

There is 12 remainder.

Each child gets 12 oranges, and there are 12 left over. This is right, for 12 multiplied by 29 = 348, and 348 + 12 = 360.

*   *   *   *   *

8.

31)99,587

In No. 8, keep on dividing by 31 until you have used the 5, the 8, and the 7, and have four figures in the quotient.8.

9.　　　10.　　　11.　　　12.　　　13.

22)253　22)2895　21)88913　22)290　32)16,368

Check your results for 9, 10, 11, 12, and 13.

These are two very different methods of approach. The deductive method alone gives the real rationale: it alone shows each step as a necessary inference from our decimal system of notation. And it is sometimes thought that there is no alternative to understanding the rule in this way and not understanding it at all; that it has to be grasped either as a process wholly logical and wholly explicable, or as a meaningless ritual to be carried out with a blank unquestioning mind. But Thorndike does not think so. He thinks the inductive method gives a *via media* which not only secures a certain measure of reasoning, but also, and this is more to the point, secures the only sort of reasoning that the average child can compass. He puts it this way: "At one extreme is a minority to whom arithmetic is a series of deductions from principles; at the other extreme is a minority to whom it is a series of blind habits; between the two is the great majority, representing every gradation but centring about the type of the inductive thinker."

There is little doubt that children reason much less than we think they do. So long as they know how to do a thing they don't worry their heads about the reason why it is done in that particular way. It is enough to be told that that is the way in which it is done. They will cheerfully accept this (and very much more) on the mere *ipse dixit* of the teacher. They resemble the youth whose tutor offered to prove to him that the square on the hypotenuse of a right-angled triangle is equal to the sum of the squares on the other two sides. The youth replied that he would not trouble his tutor for the proof, but would take his word for it as a gentleman.

Mr. Benchara Branford, in his admirable book *A Study of Mathematical Education,* has given convincing proof that to a young child the very simplest mathematical axiom is not self-evident. I repeated some of the experiments suggested in Mr. Branford's book and came to the conclusion that an ordinary child reaches six or seven years of age before he really gives intellectual assent to the truth that "two things which are equal to the same third thing are equal to one another."

I am convinced that nearly all children work the common rules in arithmetic by rule of thumb. They work them by rule of thumb whether they were taught the reason or not, whether they understand the reason or not, whether they can explain the reason or not. They work them as adults work square root—by the method given in the textbooks. Indeed, adults are little better than children in these matters. Rarely can business men give a reasoned account of the rules they employ. What explanation they give is a surface explanation: it does not penetrate to the decimal scale of notation which forms the structural basis of our system. It does not deduce from first principles. And indeed men of the highest culture employ the rules of arithmetic in the same unreasoned way. Nor can they always, when challenged, render a reason. I once gave a public lecture on Arithmetic, with a bishop in the chair. During the course of the lecture I had to demonstrate on a blackboard the two current methods of working subtraction. The bishop was much interested, and confessed at the close that he had worked subtraction all his lifetime without knowing the why and the wherefore. He certainly did not know that he had employed the method of equal additions. He had duly borrowed and had duly paid back and had rested in the belief that the account was properly squared. And he was none the worse as a practical man, and none the worse as a bishop.

That Thorndike is right in his main thesis there can be little doubt. People use the rules of arithmetic as arbitrary modes of procedure: they do not indulge in the luxury of explaining them. This does not mean that reasoning is absent: it means that it is engaged in other aspects of the general question—in the application of the rules and in the sensibleness of the result. Everybody uses reason when he is compelled to; but he doesn't waste it. In Thorndike's inductive method the very starting-point is a problem—a pressing invitation to think. The way to reach a practical solution is shown and the reasonableness of the answer is demonstrated. The only reasoning that is left out is the reasoning that nobody employs. Nobody, that is, except a mathematician; nobody except the real student who wishes to explore the whole field and leave no unexplored territory behind. And even he only brings it in when he wishes to develop arithmetic as a pure science as distinct from an applied science.

It is here that we touch the weak spot in Thorndike's armour. He has convinced us (if we needed convincing) that most of our pupils learn arithmetic by inductive methods; that the reasoning they use is the least that will serve their purpose; and that they get little profit from deductive logic. We are persuaded that this is how they begin; but we are not persuaded that this is how they end. In point of fact their interest in rigid reasoning increases as they grow older. Not only does their capacity to reason in general grow with their growth, but their capacity to reason in arithmetic grows with their practical knowledge of arithmetic. So there comes a time when they can be made to examine the machinery which they have already used, to pull it to pieces and see how the parts fit into a logical whole. I well remember the thrill with which I discovered that there was reason behind the borrowing business in subtraction. I must have been fourteen, or even older, when I found it out. I don't suppose it made me work subtraction any better than before; but it certainly increased my respect for arithmetic as a logical science— as a web of reasoning each strand of which would bear the strain of the most severe scrutiny.

Mr. Bertrand Russell, in his book *On Education*, writes as follows: "I remember a sense almost of intoxication when I first read Newton's deduction of Kepler's Second Law from the law of gravitation." Although he thinks that "logical accuracy is a late acquisition, which should not be forced upon young children," he does not believe in omitting the logical accuracy. "Rules must be learnt, but at some stage the reason for them must be made clear; if this is not done, mathematics has little educative value."

There is, again, the minority admitted to be capable of deductive thinking from the first. It is a small minority, perhaps a very small minority; but its worth cannot be estimated by counting. And to ignore its claims in the interests of the mediocre is to sacrifice the "seven men that can render a reason" to the seven hundred that cannot.

The conclusion of the matter is this. The wise teacher will refuse to make a clean-cut choice between deductive and inductive methods: he will use both. While putting his trust in the inductive method as his main stand-by, he will not withhold deductive explanations from those who can follow them, nor will he forget that as his pupils grow older they grow wiser, and are often able to understand at fourteen arguments which they were quite incapable of understanding at ten.

CHAPTER IV

MOTIVATION

With heads bent o'er their toil, they languidly
Their lives to some unmeaning taskwork give.
> MATTHEW ARNOLD: *A Summer Night.*

My life is one dem'd horrid grind.
> MR. MANTALINI in *Nicholas Nickleby.*

ARITHMETIC may be an exhilarating exercise, or it may become a nightmare. In Victorian days it became a nightmare. In elementary schools, that is. In grammar schools it was regarded lightly, though not light-heartedly; but in the elementary schools of the seventies and eighties it was a terribly serious business. For one dark period, even the teacher's salary partly depended on his success in teaching arithmetic: he was paid according to output. And for thirty years or more every child was examined in arithmetic by Her Majesty's Inspector once a year. Once a year, with the regularity of the seasons, he came round and imposed on each child in each standard a test of four sums—no more and no less—three "rule" sums and a problem. The three rules for Standard IV, the class that seems to have borne the biggest burden of figures, were reduction, multiplication of money, and division of money. Here are examples of rule tests actually set:

(1) Reduce 849,612 drams to cwts.

(2) £26 15s. 7¾d. × 278.

(3) £416,073 12s. 7½d. ÷ 381.

The test changed but little from year to year. The type of sum was fixed and the magnitude of the numbers could be approximately foretold. The examination was accordingly rehearsed throughout the year. It was rehearsed with assiduity and much tribulation. And the helpfulness of habit was by no means overlooked. The four sums were assigned their proper places on the paper, and kept there. In the lower standards it was slates, and the four corners of the slates seemed specially designed for the four sums. So each sum was given its own corner, as was seemly and proper. And by dint of much practice an extraordinary facility and accuracy were secured. The goal was absolute accuracy in the three mechanical sums, and a sporting try at the problem. Since two sums correct out of the four secured a pass, the problem didn't much matter. For success in that the teachers trusted to Providence, which, as Samuel Butler has assured us, means that they would chance it. But much time was spent over the mechanical grind—often two lessons per day with a spell of homework. A class of Standard II children, children from eight to nine years of age, could at the end of the year multiply such numbers as 48,967 × 798 with astonishing accuracy; but when asked to multiply 24 by 24 they would be completely nonplussed. When the numbers were put down, they didn't look in the least like a multiplication sum, nor indeed like any other sort of sum that had come within the children's ken. There was little thought in the arithmetic of those days, and little joy; but in its own narrow sphere it had merits not to be despised. Indeed, the goal of absolute accuracy in the "rule" sums was reached all right, though the road was thorny and watered with many tears.

Then came the emancipation. By the middle of the nineties the annual examinations had entirely ceased, and teachers were free to devise their own schemes of study and to pursue their own modes of teaching. To reform the arithmetic was one of their first cares. It was clear that it needed leavening and sweetening. It must be made attractive to the ordinary little creature of flesh and blood. But how? Some put their trust in concrete numbers. They declared the remedy to lie in discarding all abstract quantities and making the quantities real. It was believed that children hated multiplying 5247 by 78, but if they were allowed to write the word "apples," or "elephants," or "bales of cotton" after the 5247, they would regard the multiplying as a privilege and a joy. It was weariness to the flesh to reduce 5 hundredweights to ounces, but to reduce 5 hundredweights of coal to ounces was as cheery a business as reducing them to ashes.

But it was not long before the hope of conquest by concretion was irrevocably crushed. Not only did it break down in practice through failing to kindle the interest of the children, but it proved very doubtful as theory. It was found impossible in fact to leave the abstract out of the reckoning. When 5247, in the example above, is turned into apples, the 78 cannot be turned into anything but times, which is still abstract. Indeed, it is a tenable theory that arithmetic is in its very nature an abstract science, and that all its operations are performed with abstract numbers, and with abstract numbers only. The concrete interpretations we may give them are no part of the arithmetic proper. The arithmetic may stay the same while the meanings masquerade in a variety of types and trappings. $7 \times 4 = 28$ is a mathematical statement which tells us how many days there are in 4 weeks, or how many farthings there are in 7 pennies, or how many buns there are in 7 bags with 4 buns in each bag, or how many square inches there are in a rectangle 7 inches long and 4 inches wide. The actual story it tells depends upon the quest or the question. Thorndike holds this view strongly and formulates it thus:

"In all computations and operations in arithmetic, all numbers are essentially abstract and should be so treated. They are concrete only in the thought-process that attends the operation and interprets the result." [1]

He points out that we do not hesitate in algebra to divest a number of its reference to actual things. We do not let x equal the horses: we let x equal the *number* of horses, and then drop the idea of horses out of our consideration. Arithmetic should be disencumbered in the same way. "Addends must be of the same denomination," says the traditional rule, "and the sum the same as the addends." Thorndike's comment on this

[1] *The Psychology of Arithmetic,* p. 88.

takes the form of a story.[2] Several classes in a Normal College were given this problem to solve:

"In a garden on the summit are as many cabbage-heads as the total number of ladies and gentlemen in this class. How many cabbage-heads in the garden?"

And the solution looked like this each time:

> 29 ladies
> 15 gentlemen
> 44 cabbage-heads.

Sir Oliver Lodge holds the diametrically opposite view. To him everything is concrete. Even "the symbols of algebra are concrete or real physical quantities, not symbols for numbers alone."[3] He makes no bones about saying that 6 feet × 3 feet gives 18 square feet, and holds that $\frac{60}{1760}$ yards might represent the number of telegraph posts per mile. He even goes so far as to present the following ratio:

$$\frac{330 \text{ yards} \times 16 \text{ square yards} \times 77 \text{ lb.}}{4 \text{ inches} \times \frac{1}{2} \text{ mile} \times .14 \text{ ton} \times 5 \text{ minutes}'}$$

and to say that to the experienced eye it represents a velocity.[4]

---

[2] *Idem*, p. 87.

[3] *Easy Mathmatics*, pp. 53-55.

[4] "I once had difficulty in persuading another of my betters that if you repeat five shillings as often as there are hairs in a horse's tail, you do not multiply five shillings by a horsetail."—A. DE MORGAN: *A Budget of Paradoxes*, bk. ii, p. 251.

Whether Sir Oliver Lodge is right or wrong (personally I side with Thorndike in this matter), one thing is certain: the boy at his lesson is no happier in dealing with concrete numbers than with abstract numbers. In fact, concretism in teaching fails of its avowed purpose.

So does the second proposal—the proposal that all questions in arithmetic should be put in the form of problems. The problem is the cure for apathy and reluctance. Do not all children love puzzles? And are not problems puzzles? All we have to do, then, is to boycott the mechanical sum and let our youngsters wallow in problems. Unfortunately they refuse to wallow. They find the problem no more to their liking than the mechanical sum. Indeed, less so. Let the reader ask a number of school-children which they prefer, a straightforward sum (children always call it that) or a problem, and he will find that two out of every three will say they prefer the straightforward sum.

There is, I fear, no help for it. We must face the cold fact that arithmetic, however much it is doctored or dressed up, is not an interesting subject to the ordinary young child. It is true that we occasionally find children at a Montessori school who suddenly manifest a strange passion for counting or for working sums. They will diligently count up to thousands, or work sums with extraordinary zest for a whole morning. But the fire soon burns itself away. It is more like a craze for a new toy than an appetite that grows by what it feeds on; a passing fancy, not a permanent love.

It is not that the subject lacks its domains of pleasure, but that the approaches to those domains are dense with difficulties. Much may be done by a wise manipulation of the difficulties. They should be "spread out thin," as Mr. Bertrand Russell puts it, or, as Descartes counselled long ago, they should be divided up and tackled one at a time. Yet even this is not enough. A taste for the attack has in some way or other to be engendered.

Fortunately there is one thing that all pupils like, and that is success. There is one joy that never palls—the joy of achievement. And one means of bringing that joy into the arithmetic lesson is to capture some of the spirit of sport. The keenness of the sportsman can cope with all forms of drudgery and turn a task into a delight. No toil is too severe for the golfer bent on reducing his handicap. He will trudge about all day under a hot sun, hitting a ball with a stick, then searching for it, then hitting it again and repeating his search; and he comes home tired and happy and talkative. He gets no pay for his toil. Indeed, he pays for it as a privilege, pays as Tom Sawyer's companions paid for whitewashing a fence. It is sport, and he enjoys it hugely. He has not only the pleasure of beating the other man (that does not always happen), but he has the pleasure of beating the perversity of things, and, it may be, the pleasure of beating his own record. What counts is not the vulgar joy of getting the better of other people, but the nobler joy of getting the better of himself and his difficulties. To taunt this fine zest and gusto with its lowly birth—to trace its origin to such instincts as pugnacity and self-assertion—is merely to charge it with being common to mankind. The instinct may freely be admitted; but it is instinct rightly directed, instinct sublimated, instinct operating in a sphere superbly human.

Children, too, in the playground and the playing-field, display this fine spirit almost spontaneously. Team strives with team. Each player plays for his side instead of for himself. It is his team that competes: he himself co-operates. And while victory brings delight, defeat brings neither rancour nor despondency: it merely stimulates to greater effort. And thus our children not only learn to play the game: they learn the game itself. And if our boys and girls can be brought to feel that arithmetic is a game—a noble game—one of the noblest though not one of the most spectacular that the human race has played—and that it is an honour and a privilege to play at it; and if we can keep that feeling alive by the right exercise and the apt stimulus, cunningly applied with a smile and a jest, as becomes so noble a game, the arithmetic lesson will cease to be a dismal grind and become a grand pursuit full of glamour and excitement.

Utopian! you will say. Nay, I have seen arithmetic pursued in this spirit, and with astonishing results. I say "pursued," not "taught"; for arithmetic is essentially an active subject, and the best teacher of arithmetic is not the most eloquent exponent of its principles, but rather the most skilful organiser of arithmetical studies. It means much foresight, much vigilance, much care. It means an easy grading of the work; a prompt and scrupulous marking of the sums; a keeping of records of many kinds; a display of diagrams showing class progress and team progress; the making and keeping of a private graph by each pupil for his own behoof; a nod of approval to the diligent; a word of encouragement to the struggling; a timely rebuke to the slack; an occasional team race; a frequent time test; an exhibition of work of special merit; ample opportunity for individual initiative and for forging ahead—these expedients, or indeed any other device for keeping the loins girt, the sword bright, and the lamp burning.

Since much of a pupil's good work springs from a desire to stand well with his teachers and his classmates, care should be taken to see that his work is neither ignored altogether nor looked at with indifference. If he finds that nobody cares whether he does well or ill, he will soon cease to care himself. Other motives, it is true, operate in the classroom—motives of a higher ordar—intellectual curiosity, interest in the subject itself, and a love of truth in general. But when these higher motives fail, the sporting spirit is always there as a stand-by. And to get the sporting spirit the teacher must be a sport himself.

CHAPTER V

GETTING SUMS RIGHT

But 'twas a maxim he had often tried,
That right was right, and there he would abide.

CRABBE: *The Squire and the Priest.*

I HOLD the view that children should get their sums right; and that if they don't get them right at first, they should get them right at last. I go even further, and say that if they don't get their sums right, nothing else can compensate. I mention it because I often come across people who think otherwise; people who assert that it doesn't matter whether the sums are right or wrong so long as the pupil is working intelligently, or is getting a grasp of some important principle, or is strengthening his moral fibre by grappling with difficulties, or is acquiring some other virtue which is supposed to be of a higher order than a simple quest for a definite piece of truth.

The very soul and essence of arithmetic is logical accuracy. A question is asked in abstract arithmetic. The answer is either right or wrong. There is no compromise; there can be no compromise. To blur the distinction between right and wrong, or to blunt the sharp contrast between them, is as bad in arithmetic as in morals. In morals we can make no terms with the devil. And in arithmetic getting sums wrong is the very devil. Children need not be told this—certainly not in these terms. They need not be told it, because they know it already. They love getting sums right and consider it a merit; they hate getting them wrong and consider it a lapse. At least they begin that way; and they will continue that way unless they are educated out of it. There are many ways of educating them out of it, of dulling their keen sensibility to right and wrong; and one of them is to neglect marking their sums.

A child with his arithmetical conscience as tender as it should be wants to know, as soon as he has worked a sum, whether it is right or wrong. He not only wants to know, but he cares very much whether it is right or wrong. And the longer the delay in discovering, the less he will care. Everybody is the same with a fresh product of his brain. Observe an artist dallying with a new sketch. He is supposed to have finished it, yet he cannot leave it out of his sight. He looks at its reflection in the mirror. He leans it against a chair so that he can view it while eating his lunch. He fixes it at night where he can catch a glimpse of it when he wakes in the morning. And he is keen on getting your criticism. In a day or two his interest cools down, and he will put his picture in a drawer and forget all about it. So is a boy with his sum. He wants it marked while it is hot from his brain. And it is then that the marking will do him most good; for the steps by which he reached his answer being fresh in his mind, he can correct them more readily than when they have faded from memory. So the first desideratum of marking is that it should be prompt.

The next desideratum is that incorrect sums should be immediately re-worked—re-worked by the pupil himself. There is no correction like self-correction. Correction by the teacher on the blackboard is a very poor substitute. His exposition may be most sound, but it is superfluous. For errors in arithmetic are nearly always due to carelessness, and the cure for carelessness is not logic but care. It is certainly not cured by doing for the careless ones a piece of work which they should do for themselves. Nor is anything achieved but boredom by forcing upon the whole class, careless and careful alike, an explanation which nobody needs.

An ideal system of marking is hard to devise. I don't think it can be done without delegation: the teacher must entrust much, if not most, of the marking to others. Personally, I think the proper marker is the child himself. The plea that he is not to be trusted cannot be maintained. Until we have put the matter to the test of experiment or of experience we do not know how much children can do for themselves and by themselves. Little children in nursery schools—children of three and four years of age—can perform acts of social service which most people would regard as impossible. They can dress themselves, can serve at table, can carry plates of soup long distances without spilling a drop. Instead of shirking responsibility, they welcome it. And the responsibility of marking sums is no greater than the responsibility of carrying soup. There is, of course, the temptation to cheat in the marking of sums. But a risk of this kind is worth taking, as has been discovered by public libraries which have adopted the system of open shelves. The number of books that disappear under this system is no larger than under the catalogue-and-attendant system; and the loss, in any case, is far less than the cost of the extra staff. The risk of dishonest marks is worth taking. The defaulters are few and easy to detect, and suitable penalties not difficult to devise.

A system of self-marking does not mean the absence of supervision. As the pupils get older, they have to deal with more complex examples, and the mere arrival at the right result, necessary as it is, is not enough. The route by which they arrive is often of consequence. To reach the goal by a long route when a simple short route is possible, or to neglect to show an orderly and logical sequence of steps, is to miss much of the benefit to be gained by the exercise. Hence, whilst always allowing the trustworthy pupils to have free access to the answers, the teacher should systematically examine the exercise books with a view to correcting a clumsy or illogical solution.

A child loves getting sums right. A mere tick on his paper makes him purr. This feeling is natural and proper—a thing to be cherished with assiduous care. To crush it by constantly giving him sums that are beyond his powers, or that betray him into frequent pitfalls, or that appal him by their apparent difficulty, is to destroy the teacher's most potent resource. For if the pupil loses heart he loses everything. Not that the sums should be so easy that they can be done without effort, but that they should be so easy that they can be done without a disheartening and demoralising amount of effort. I pin my belief to the slightly difficult sum—the brief and simple sum which presents its own little point of difficulty. The slightly difficult sum has, I venture to think, more real training value than the very difficult sum. It is, as I have pointed out elsewhere,[5] a question of light dumbbells versus heavy dumb-bells. It is also a personal question. What is an easy sum to one child is a difficult sum to another. Each requires something a little beyond his easy reach—something which will make him stretch without making him strain; something which will be more likely to bring the joy of overcoming difficulties than the dejection of hopeless failure. When a pupil gets, on the whole, more sums wrong than right, when, that is, his initial chance of failure is greater than his initial chance of success, we can be quite sure that he is working in the wrong spirit. He either had the wrong spirit to start with, or he gained it from the sums themselves. There are difficulties that inspire, and difficulties that discourage. The latter we must sedulously avoid.

If getting sums right in plenty is, as I believe it is, a sound practical aim in the arithmetic lesson, a large portion of the time should be spent in individual and independent work. Arithmetic is *par excellence* an active thing. It is a thing we *work*. We do not listen to it; we do not read it (except in the university sense of the word); we worry it out for ourselves. Mark how much easier it is to solve a problem for ourselves than to follow the

[5] *Group Tests of Intelligence,* p. 77.

steps of somebody else's solution. The light that leads to the goal comes from within, not from without. And it flickers and flashes with a rhythm which is irregular in itself and is peculiar to the person concerned. Every pupil in arithmetic has his own *tempo*. He should be allowed to go through the course at his natural pace. He should not have to wait for others to catch him up, nor should he be hustled through exercises which bewilder and depress him.

CHAPTER VI

THE KING'S HIGHWAY

The longest way round is the nearest way home.

OLD PROVERB.

Next he showed them the two by-ways, that were at the foot of the hill, where Formality and Hypocrisy lost themselves.

BUNYAN: *Pilgrim's Progress.*

ARITHMETIC is not only a science, but an art as well. It aims at doing things. And there are traditional ways of working sums, ways that represent the wisdom gained by long and varied practice. Clumsy methods, methods that wastefully use up time and energy, have been left behind, while neat, safe, and effective methods have been carefully preserved and passed on from generation to generation. So a traditional method is presumptuously a good method. The odds are heavily in its favour. And any change that is proposed should be cautiously, if not suspiciously, examined. Nay more; it should be subjected to the test of rigid experiment, and the new method rejected unless it is demonstrably better than the old. A few solid arguments in its favour are not in themselves sufficient. It is said that the translators of the

Authorised Version of the Scriptures received from a critic a suggestion that they should alter a certain word in their version. He gave five sufficient reasons for the change. They replied that they had already considered the matter and had fifteen sufficient reasons against the change.

In treating the common "rules" of arithmetic, it is well to have one good standard mode of procedure for each type of sum. There are certain clear criteria which each rule should satisfy: it should be simple, it should be easy, it should be safe, and it should be universally applicable. When I say that it should be simple and easy I am speaking relatively; what I mean is that it should have no *unnecessary* complications and no *unnecessary* difficulties. The method having been carefully and definitely chosen, it should be regarded as the King's highway which all should learn to tread before they venture upon by-paths and short cuts—the road which always *may* be travelled, and, unless there is good reason to the contrary, the road that always *should* be travelled. It is important, therefore, that this universal fairway should be a good one, that it should be known first, and that it should be known well.

A few simple examples will clarify what I have just said. When little children are first set to add numbers on paper, the figures may be arranged as in (*a*) or as in (*b*).

(*a*) 3
8
5

(*b*) 3 + 8 + 5

(a) is better then *(b)*, because *(a)* is on the king's highway, while *(b)* is not. It can be proved experimentally that children find it easier to deal with the figures in the *(a)* form than in the *(b)* form. They take less time over it and are more accurate. That alone is a good reason for beginning with *(a)* rather than with *(b)*. There is, moreover, the further and more cogent reason, that *(a)* is the general method, the only method that is universally suitable for all addition sums, difficult as well as simple. *(b)* is a by-path. It is not a bad by-path: it does not lead away from the main road; but it is not so good as the main road itself.

Let us now consider an example of addition with higher numbers, as in *(c)* and *(d)*.

$$(c)\ \begin{array}{r} 28 \\ 64 \\ \underline{37} \end{array} \qquad (d)\ 28 + 64 + 37$$

If we start with the units in both cases we shall find ourselves travelling along parallel routes, with *(c)* as the main road and *(d)* as the by-way. We are sometimes, however, advised to work *(d)* by adding the tens before the units, and to proceed thus: 28 and 60 are 88, and 4 are 92, and 30 are 122, and 7 are 129. This is the method advocated by Miss Punnett in her book *The Groundwork of Arithmetic*.[6] She not only recommends the horizontal

[6] If my own principles are well founded, I am forced to differ with Miss Punnett in some of her methods. This, however, does not prevent me from admiring the general reasonableness of her system, nor from paying a tribute to the admirable influence she has had in the teaching of arithmetic in London schools. Whenever I find a class-mistress teaching arithmetic

arrangement as preliminary to the vertical arrangement, but addition from the left as preliminary to addition from the right. But preliminary exercises are useful to the extent that they lead into and enforce the main current; when they lead away from it they are wasteful. In this instance they lead towards Hindu methods.

It is interesting to know that the Hindus, from whom we ultimately derive our system of notation, used to add the higher denominations before the lower.[7] This is how they would perform the addition sum given below:

$$
\begin{array}{ccc}
6 & 8 & 4 \\
8 & 7 & 6 \\
4 & 9 & 5 \\
9 & 5 & 7 \\
\hline
\end{array}
$$

```
     2  7  9  2
  3  9  1
     0
```

Arguing (quite validly) that the largest denomination is the most important, they began with the hundreds column. Finding it come to 27, they put the 27 down. They then added the tens column, putting the 9 of the 29 in the tens column and changing the 7 to 9. This they could do more easily than we because their writing materials permitted of the ready erasure of figures. The units column adding to 22 necessitated three further changes in the figures already written.

with enthusiasm and vigour, I almost invariably find that she has been inspired by Miss Punnett.

[7] See Cajori's *History of Elementary Mathematics*, p. 96.

They had a very cogent reason for working from left to right; but it has been overriden by still more cogent reasons for working from right to left. It is practical expedience that has determined the survival of the present method.

For every "rule" there should be a general all-inclusive method—a method that admits of no exceptions. It follows that short division can never be a general method for division. Although it is possible to divide by some composite numbers by resolving them into factors less than 13, it is not possible to treat all composite numbers in that way; and as for prime numbers over eleven, they are all left out in the cold. There is, in fact, only one general rule for division, and that is long division. Division by factors should not be taught at all until the pupils are familiar with the universal highway. And then it will be found unnecessary.

Short cuts are sometimes useful; but rules for short cuts are often abominable. There is one rule for multiplying by 25 and another for dividing by 25. I can never myself, by brute memory, recall which is which, but have to revert to the fact that $25 = \dfrac{100}{4}$ and deduce the rule from that. The same is true of multiplying or dividing by 125, i.e. by $\dfrac{1000}{8}$.

The rules for multiplying by dozens and by scores, useful as they sometimes are, do not figure so largely in our daily lives as they do in the mental arithmetic lesson. We rarely buy pounds of butter by the dozen, nor do we often buy penknives by the score. When a magnitude is just short of a round number, such as 99 or 5 ft. 11 ins. or 3*s.* 11¾*d.*, it is obviously easier to deal with the round number and subtract the deficit. This device is, however, frequently overdone. If I set a class of bright children to multiply a sum of money by 193, some will take 7 times the sum from 200 times the sum. From one point of view it is an intelligent method; but if intelligence should aim at getting the correct answer in the readiest way, the method can only be described as stupid. For the children in question almost invariably come in last and bring the wrong answer. There is much more to be said for adopting the subtraction device in multiplying an abstract number by 193 than in multiplying compound quantities; but even then the liability to error is greater than when the trodden path is followed.

A textbook on short methods, however good it is, is generally padded out with rules like this:

To multiply any two numbers under 100 when the sum of the tens equals 10 and the units are alike.

Rule.—To the product of the tens add the common unit, call the sum hundreds, and annex the product of the units.

Examples.—76 × 36 and 67 × 47

(*a*) 76
36
―――
2736

(*b*) 67
47
―――
3149

(a) $7 \times 3 + 6 = 27$ (hundreds) and $6 \times 6 = 36$

(b) $4 \times 6 + 7 = 31$ (hundreds) and $7 \times 7 = 49$

Very interesting, but quite useless. Who can remember a rule like that? Who wants to remember it? To multiply the numbers out in good honest fashion is a far less strain on the mind, and gives one a much greater feeling of confidence in the answer.

Occasionally, of course, short cuts are safe and expeditious, as for instance when use is made of the fact that if a fraction is nearly a whole number, multiplication may be made easier by subtracting the deficit. For example, we can find ¾ of 256 by putting down 256, writing underneath a quarter of it, i.e. 64, and thus subtracting—a much shorter method than the usual one of multiplying by 3 and dividing by 4. In the same way 3⅞ may best be regarded as 4 – ⅛.

This is an example of some of the simple and safe by-paths which are visible from the main road, and which indeed can be found only by those who have already travelled along the main road. It follows that the teacher's cardinal care will be to see that his small band of adventurers in the realm of arithmetic may be made thoroughly familiar with the highways before they begin to explore the byways.

CHAPTER VII

THE BEGINNINGS

The arithmetic of babes.

WORDSWORTH.

IN teaching numbers there is little doubt that we begin well. A young child in the infant school or the kindergarten gains his first notions of number through his eyes and his fingers. He handles real things. He counts beans and beads and tablets; he performs simple operations with them; he adds them and subtracts them; he arranges them in groups and disposes them in patterns; he builds up the multiplication table with them and makes them answer questions printed on a card or asked him by the teacher. The consequence is that his concepts of the simpler numbers and of their relationships are singularly clear and accurate. The meanings he acquires are real meanings, gained by a living experience. And his knowledge of number, being intelligently gained, can, when need arises, be intelligently applied. In fine, the foundations of arithmetic in a good modern school are well and truly laid.

It was not always so. I well remember the time when the teaching in the infant school was indistinguishable in kind from that of the senior school; and until fifteen years ago the individual study of number was unknown. The children were always taught *en masse*. When concrete objects were used, the number lesson was just like the drill lesson. The children were asked to do this, and they did this; to do that, and they did that. They obeyed mechanically and collectively. They were invited to think of what they were doing; and they were supposed to think abreast just as they were able to march abreast. Yet there is no time which the flow of thought is more manifestly fitful and intermittent than when it deals with a problem in arithmetic—and to the young beginner every exercise in number is a problem: it presents a difficulty to be overcome by an act of thought rather than by an act of memory. Hence the children's mode of attack in our most enlightened schools is individual and personal.

I do not propose to deal in detail with the number work done in the infant school, partly because I think it is fundamentally right in its general trend, and partly because the particular methods are in the main experimental and tentative. It is difficult as yet to say which of the methods are most effective. Besides, others have dealt with the beginning of arithmetic in a way which I cannot hope to improve upon. First of all there is Miss Margaret Drummond's little book, *The Psychology and Teaching of Number,* which embodies the results of much keen and patient observation of the child's response to simple arithmetical questions, and offers much wise counsel to teachers of the very young. Miss Punnett's well-known book *The Groundwork of Number,* and Miss Mackinder's *Individual Work in Infants' Schools* should also be read by all who undertake to teach number to young children.

Miss Drummond reaches one conclusion which I regard, on quite independent evidence, as profoundly true. It is that in teaching number we cannot force the pace. In her own words, "The knowledge of number and the ability to perform number operations that is acquired in the Infants' Department (ages 5 to 7) are mainly the result of mental growth; this growth takes place, not because of the teaching, but often rather in spite of it." When the dedicatee of this book had reached her third birthday, I tried to develop in her mind the notion of three. But I failed. She knew the story of the three bears; she could count "one, two, three" (only she would insist on adding "Go!"), but she could not tell me how many pennies she had in her hand when I put three there. A month or two later I found she had acquired for herself the very notion which I had laboriously and fruitlessly tried to teach her. I am quite willing to accept Miss Drummond's dark suggestion that she would have acquired it sooner if I hadn't tried. Often have I in past years listened to teachers giving most excellent lessons on the analysis of number to young children and have marvelled at the meagreness of the results. To a listening visitor the lessons seemed capable of illuminating the mind of a mollusc, and yet the children seemed at the end of the lesson just where they were at the beginning.

On the next page I reproduce from *The New Examiner* my one-minute number tests, which I have used extensively for the last fifteen years or so. Each child is examined individually and in isolation. He is asked the question, "One and two?" and as soon as he answers it he is asked the next, "Four and one?" and so on. He is not allowed to proceed until he has given the right answer. The examiner repeats the question, but gives the child no help of any kind. From the last question answered within the minute the arithmetical age may be read off on the scale. Thus a child who answers 12 addition questions in a minute has an addition age of 7 years 6 months, while if he answers 12 subtraction sums his subtraction age is 8 years and 3 months.

My purpose in introducing the test here is to illustrate and enforce my general thesis that in the early stages of arithmetic the pace cannot be forced. The scale was carefully constructed about the year 1914 on the basis of results obtained by myself and others in the application of the test to several thousands of boys and girls. Since the scale was devised, new methods of teaching have gradually crept into the school. In nearly all schools the early number work is now individual, and in many schools it is spontaneously undertaken by the children. They receive no formal lessons, and sometimes indeed are allowed to neglect number for days and weeks together. The teachers no longer strain and strive to impart new numerical ideas. And yet the average number of items correct at a given age is the same to-day as it was fifteen years ago. The leisurely

One-minute Oral Addition Test			One-minute Oral Subtraction Test		
Question.	Addition Age.		Question.	Subtraction Age.	
	Yrs.	Mths.		Yrs.	Mths.
(1) 1 + 2			(1) 2 − 1		
(2) 4 + 1	5	0	(2) 3 − 2	5	5
(3) 2 + 2	5	3	(3) 5 − 1	5	9
(4) 2 + 4	5	6	(4) 6 − 2	6	0
(5) 3 + 2	5	9	(5) 5 − 3	6	3
(6) 4 + 3	6	0	(6) 2 − 2	6	7
(7) 2 + 5	6	3	(7) 7 − 2	6	10
(8) 5 + 4	6	6	(8) 6 − 4	7	2
(9) 3 + 5	6	9	(9) 7 − 3	7	5
(10) 8 + 2	7	0	(10) 6 − 3	7	9
(11) 4 + 4	7	3	(11) 8 − 2	8	0
(12) 5 + 2	7	6	(12) 7 − 5	8	3
(13) 6 + 4	7	9	(13) 8 − 3	8	7
(14) 1 + 8	8	0	(14) 7 − 4	8	10
(15) 3 + 7	8	3	(15) 9 − 3	9	2
(16) 6 + 3	8	6	(16) 8 − 5	9	5
(17) 2 + 6	8	9	(17) 10 − 4	9	9
(18) 5 + 5	9	0	(18) 9 − 5	10	0
(19) 7 + 2	9	3	(19) 10 − 3	10	3
(20) 4 + 6	9	6	(20) 9 − 4	10	7
(21) 7 + 5	9	9	(21) 11 − 2	10	10
(22) 8 + 3	10	0	(22) 10 − 6	11	2
(23) 4 + 9	10	3	(23) 12 − 3	11	5
(24) 6 + 8	10	6	(24) 11 − 6	11	9
(25) 7 + 6	10	9	(25) 12 − 5	12	0
(26) 9 + 8	11	0	(26) 13 − 4	12	3
(27) 9 + 6	11	3	(27) 15 − 9	12	7
(28) 8 + 7	11	6	(28) 14 − 6	12	10
(29) 5 + 9	11	9	(29) 17 − 8	13	2
(30) 7 + 9	12	0	(30) 16 − 7	13	5

methods of to-day are just as effective as the strenuous methods of the past.

It is worthy of note that while modern individual methods in arithmetic have made no difference to the arithmetic, they have made a marked difference to the reading. Judged by a similar test for reading, children of a given age are apparently six months in advance of what they were fifteen years ago.

How are we to begin teaching arithmetic? The answer is clear and unambiguous. The first thing a child must learn is counting. Until he can count he can do nothing else. The whole fabric of arithmetic rests on the natural number series, and every arithmetical process can in the last resort be reduced to counting. All the "rules" of arithmetic are but expedients for shortening the time and labour of counting; and the results we arrive at tell us no more than we could discover by counting: they only tell it more quickly. We must therefore look upon counting as the ultimate logical ground and justification for any process, rule, or method we may employ. Miss Drummond rightly contends that no child should receive formal lessons in arithmetic until he can count things accurately. She even goes so far as to say that until this power is acquired, formal lessons do positive harm.[8]

It is by no means easy to say when a very young child can really count, for there is such a thing as spurious counting. It does not follow that because he can say "one, two, three, four" he means by those words

[8] *The Psychology and Teaching of Number,* p. 119.

51

what you and I mean by them. They may be to him mere rhythmic sounds, as empty of arithmetical meaning as "Fee fi fo fum." He descends the stairs, saying, "One, two, three," etc., as he steps from stair to stair. But the word "three" is just as likely to mean "third" as "three"; and just as likely to mean nothing as either. It is only by close questioning—by asking him to fetch you four books from a shelf, or to tell you how many corners a door has, or how many fingers there are on one hand with the thumb left out—that we can be quite sure that he can count. And when he can count accurately and intelligently he is ready to forge ahead at great speed.

Just as younger children don't count when they think they do, so older children count when they think they don't. Observe them "counting out," before beginning a game, by means of the jingle "ena dena dina do." Tell them that they can achieve exactly the same end by counting up to 29 (the number of accents in the jingle) and they will be surprised and incredulous. The end of a count is inexorable and predictable; but Heaven only knows on whom the final syllable of "ena dena" will fall. It's not counting, it's magic.

The four fundamental processes in arithmetic are merely four different ways of counting. Adding is counting forwards, and subtraction counting backwards. In multiplication or division we count forwards or backwards by leaps of uniform length. In the Rudolf Steiner schools, where much importance is attached to rhythm, multiplication is regarded as rhythmical counting, and the three times table is taught thus: 1 2 3′ 4 5 6′ 7 8 9′ 10 11 12′, and so on. For the earlier tables the method seems plausible enough; but when the twelve times is reached the rhythm will, I fancy, have become a little difficult to catch.

Number has not only a serial meaning: it has a group meaning as well. It is true that the number 6, for instance, has a definite position in the natural number series (it stands between 5 and 7); but it has not, like a Euclidian point, position only. It has magnitude as well. It means a group of six objects, which is larger than a group of five similar objects, and smaller than a group of seven similar objects. And a child's knowledge of the group is not complete when he has counted it: it has only just begun. He has to envisage it as a group of a certain aggregate size, though not of a fixed pattern; he has to analyse it and see that it may be resolved into a number of smaller groups, such as 1 and 5, or 2 and 4, or 2 and 2 and 2. And after splitting up the group into smaller groups he has to put these smaller groups together again to see that they re-form the original group. Wherever the group idea is dominant, the teaching of number means analysis and synthesis. And this is exactly what it meant twenty years ago. Each of the natural numbers up to ten was dealt with one at a time and each was analysed exhaustively before the next was tackled. Everything was supposed to be learnt about 5 before anything was learnt about 6, everything about 6 before anything about 10. The first lesson was on 1—a lesson unspeakably artificial and futile. For no child can give a notion of 1 before he gains a notion of 2. One-ness and twoness (or possibly one-ness and more-than-one-ness) are twin ideas: they are born together in the child's mind. Even a lesson on two, which meant a lesson devoted wholly and exclusively to two, is not cheerful to contemplate. But it was attempted. An orderly and painstaking progression through the series, up to 10 at least, was regarded as the best way; and indeed the only way. We have now changed our views, and our tactics—not that we disbelieve in the importance of analysis, but that we disbelieve in teaching it collectively. We believe rather in leaving each child to make his own analyses.

There is yet another notion of number which has to be developed in the beginner's mind—the notion of ratio—the notion that twelve means twelve times whatever is one. If one is an inch, then twelve is a foot; if one is a penny, then twelve is a shilling. The importance of this aspect of number was first put forward with clearness and emphasis over thirty years ago in a book called *The Psychology of Number and its Applications to Methods of Teaching,* by McLellan and Dewey. The principle of ratio underlies all measuring and weighing. We can count discrete objects only; continuous quantities we have to measure. We can count separate pieces of wood, but we cannot count one long piece of timber. We take a convenient length, such as a foot, as a unit, and find how many times a foot-rule must be laid down to cover the whole length. In other words, we find the ratio between the foot-rule and the piece of timber. If it is 1: 7½, we say the plank is 7½ feet long. It is by ratio that we measure milk; it is by ratio that we weigh out pounds of sugar. It is the notion of ratio that gave rise to Tillick's bricks, and is believed to be imparted by the Montessori steps. The importance of ratio may readily be conceded. When, however, it is proclaimed as the root idea of numbers, and a whole system of arithmetic built upon it, it is time to raise a protest. Fundamental as ratio is, counting is still more fundamental. How can we find the ratio between the foot-rule and the piece of timber if we cannot count? Besides, that absolute accuracy which it is the crowning glory of arithmetic to illustrate and exemplify is not to be found outside the realm of abstract number. Once we begin to apply number to the world of real concrete things and measure continuous quantities we get nothing but relative accuracy—nothing but approximate results. We are never quite right, but only nearly right.

Even when we have realised that a given number is a unique member of the natural number series, that it represents a group of a fixed size, and that it stands for a ratio between two quantities, we cannot yet, it seems, be said to know the number completely. Thorndike distinguishes a fourth meaning, which he calls the "nucleus of facts" or relational meaning. To know the number "three" in this sense is to know not only its group implications—its internal relationships so to speak—but its external relationships as well; to know that it is 5 less than 8 and 7 less than 10; to know that it is a half of 6 and a fifth of 15; to know that it is the square root of 9 and the cube root of 27. The difficulty about these relationships is that they are inexhaustible. In analysing a number considered as a group we know when we have finished; but in dealing with its relations to all other numbers we are engaged upon an endless task.

Leaving out of account the general and radical change of method in infant-school work, and confining our attention to the teaching of number, we find the cardinal defect of the old infant school to lie in its emphasis on one aspect only (the group aspect), and the cardinal merit of the new infant school to lie in its catholicity—its inclusion on equal terms of all four aspects. The younger children are certainly allowed to count; nay, are encouraged to count. When they are attacked, as they sometimes are, by a fit of counting—a mild and healthy form of comptomania—they are given a loose rein. They may be seen laying a long train of Montessori beads, stringed in tens, all along the floor of the hall and half-way down the corridor. They are counting thousands. They feel they are doing big things. And so they are. And they not only count, they analyse and they measure. Their occupations, and the didactic apparatus which they use, force them to find relations, both internal and external, between the simpler numbers; and their exercises in weighing, measuring, and comparing develop a mental background from which a clear notion of ratio may later on emerge.

We now come to a question which has given rise to much controversy in Germany, England, and America— the question of number patterns. If a number has to be regarded as a group, it seems reasonable and helpful to have a definite mental image of that group. To put it specifically, when we think of 5 in visual terms, are we to picture it like this • • • • •, or like this; $\begin{smallmatrix} \bullet & & \bullet \\ \bullet & \bullet & \end{smallmatrix}$, or like this $\begin{smallmatrix} \bullet & & \bullet \\ & \bullet & \\ \bullet & & \bullet \end{smallmatrix}$?

It is not quite certain that it is wise to picture it like any one of them—to have any fixed image at all. It is perhaps better to vary the pattern, and even the elements of which the pattern is composed. Miss Punnett thinks so. She would have children, when dealing with the number 5, make as many designs as possible out of 5 sticks, or 5 discs, or 5 square tablets. Variety should be the aim, not fixity. Without variety it would be difficult for the child to get the abstract idea of 5 as independent of all material and all pattern. The child who can think of 5 in one pattern only, if such a child exists, is as badly equipped mentally as the child who can add apples but can't add oranges, or as the child who refused to believe that 6 and 4 made 10 because he already knew that 5 and 5 made 10. Those, however, who urge the adoption of a fixed series of patterns do not urge it as a logical necessity, but as an expedient for facile calculation.

Another kind of doubt springs up in the minds of those who have read Sir Francis Galton's *Inquiries into Human Faculty,* where it is shown that many people have fixed modes of picturing numbers in their minds, modes which are personal and peculiar to themselves. Has each child his own way of picturing numbers, a way which is for him better than any we can force upon his notice?

There can, however, be no doubt that pictorial representation is useful in the earlier stages, even if it is only a stepping-stone towards a more general or more personal concept; and if there is to be pictorial representation there is convenience in making it systematic and uniform throughout the school. What scheme of number patterns shall we use? There are several to be found in our schools. The element of the pattern is everywhere the same—a dot—but the arrangement of the dots varies from school to school and from country to country. All the systems, however, can be reduced to three. The distinguishing factor is the unit of design. In Germany and America the favourite unit is two, in England three. In all countries however, the minority of teachers affect the domino systems, when the unit is five.

LAY'S SYSTEM, BASED ON 2.

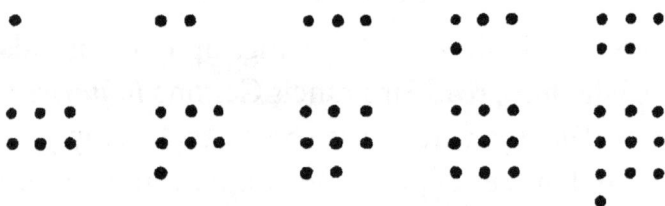

THE ENGLISH SYSTEM, BASED ON 3.

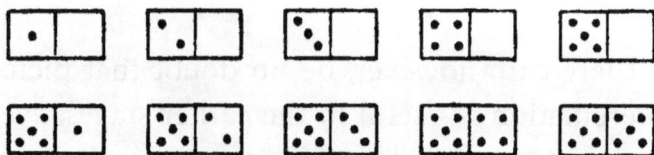

THE DOMINO SYSTEM, BASED ON 5.

Miss Mackinder advocates the three system on the ground that three constitutes the largest group that children can recognise as a unit. This is perhaps true if the dots are supposed to lie in one line, but it is not true if they form a familiar pattern. Thus four dots arranged as on a domino are quite as easily recognised at a glance as three dots arranged in a row. Indeed, experiments with the tachistoscope (a piece of apparatus which enables us to expose a number word or picture for a fraction of a second) have shown that in the matter of speed and accuracy of recognition there is not much to choose between the three systems. Personally I prefer the domino system. It seems to me to fall in more naturally with our decimal system of notation, as it makes the half of ten the unit of design, and ten itself a sort of double unit. Moreover, it accords better than the others with the arrangement of pips on playing cards (I am not quite sure whether this is an argument for it or against it); and finally (a genuine advantage this) it can be seen at a glance by how much the numbers fall short of ten. My preference for this system is fortified by the fact that I have seen it employed with extraordinary success. It has, however, one slight defect: the three cannot be converted into four without rubbing out a dot.

It is idle to pretend that we can dogmatise over the question of number patterns. We don't know enough about them. There is here a vast field for patient research. Until that field is explored, every candid account of number patterns will probably be as vague and inconclusive as this one.

CHAPTER VIII

NOTATION

We may be well assured that our system of calculation will not always be cramped by counting in one way and measuring in another.

AUGUSTUS DE MORGAN.

IT takes time (and teaching) for a pupil to realise the beauty and ingenuity of our mode of recording number. The realisation does not come of itself: it has to be evoked. The ordinary man takes our numerical system for granted. He never thinks of asking whether such a number as 268 could be expressed in any other way. He regards the whole scheme of numerical notation as part of the nature of things—an inexorable fact like the Atlantic or the British Constitution—a thing which is because it is. Yet the Romans had it not; neither had the Greeks. They calculated in their heads; they counted on their fingers or their toes, or any other computable things that were available. But they couldn't work sums as we work them: their number system did not permit them. Let us try to multiply cccxlix by cxviii by using Roman numerals only and forgetting for the moment what we know about place value. The attempt will bring home to us the colossal debt we owe, directly to the Arabs, indirectly to the Hindus, for a system of notation which alone made possible the vast structure of modern arithmetical science.

We begin to teach notation in the infant school, and all too often drop it afterwards. There we show the children the structure of such numbers as 15, of 78, or even of 135. We exhibit the device of grouping smaller units to form higher units, and grouping them according to a fixed law. We give them sticks tied into bundles of ascending order of magnitude. They therefore enter the senior school with some sense, however vague, of the general build of a numerical expression—a sense which has to be made keener and clearer when they have to extend the decimal system below unity, again when they try to understand an algebraic expression, and yet again when they enter upon the study of logarithms.

In the meantime a more practical issue demands our attention. We have to teach our children to read numbers and to write them. That this, as a simple practical art, is neglected in our schools is abundantly shown in the Report of the Board of Education on the *Teaching of Arithmetic in Elementary Schools.* There it is recorded that only 62 percent of children in Standard V can take one from 10,000 accurately. It is not to be supposed that they failed to subtract (unless they had learnt by the method of decomposition), but that they failed to put down ten thousand. This is confirmed by the result of the next test, which was: Write in figures ten thousand and ten. The percentage of successes was 66. It can hardly be claimed that these results are satisfactory. The cause of the weakness is not far to seek. High numbers have of recent years been out of fashion, and on quite plausible grounds. It is desired to bring school arithmetic into closer relation with out-of-school arithmetic, and since most of our transactions in daily life have to do with small quantities and small sums of money, the school exercises should fall into line with them. But once we pass from our own personal concerns into municipal and national concerns we get to new orders of magnitude. Poor as we are personally, we at least hear of thousands and even millions of pounds. And physical science brings to our notice numbers of a still higher order. So it is not unreasonable to expect our oldest children to be able to read and to record numbers up to a hundred million at least. Especially as it is an art quite easy to learn.

There are, however, small points of difficulty. At the earliest stage children are prone to confuse fifteen with fifty, sixteen with sixty, and so forth. If they can put down without hesitation such numbers as seven hundred and four, seven hundred and fourteen, and seven hundred and forty, their main difficulties are over.

Then comes punctuation. We punctuate numbers, just as we punctuate sentences, and for the same reason—to help the reader to take in the meaning. When figures stand in a row, the largest number that the eye can comfortably seize at a glance is three; so when the number is high we punctuate it so as to form groups of three. It is thus that the figures are presented in the most convenient eyefuls.

There is another reason for selecting three as the group: each three has the same internal nomenclature—each has hundreds, tens, and units—and each has its own group name. Look at the number 527,527,527. The last triad (or trio, or triolet, or triplet, or what you will) is a plain five hundred and twenty-seven. The middle triad is also five hundred and twenty-seven, but the unit is no longer one but a thousand. The first triad resembles the others in all respects but one: its group name is a million. What the octave is in music, the triad is in numeration: it may be repeated indefinitely at higher levels. This triadic arrangement is merely a matter of terminology, but it is a terminology which must be learnt. The pupil finds it helpful to remember that in reading punctuated numbers the word million comes in at one comma and the word thousand at another.

Punctuation is purely a matter of convenience: there is no mathematical principle involved. We don't use a comma when we write the year 1928 because it is an extremely familiar set of figures. We don't punctuate decimals because nobody has yet started doing it—nobody at any rate who can lead the fashion.

To make sure that notation is known, and continues to be known, frequent tests are necessary. And they should be searching: they should test the pupils' capacity to evade pitfalls. There is no point in asking them to put down five thousand seven hundred and sixty-two. If they know anything about figures at all they will write that number correctly. But if they are asked to put down seven, and under it to put down eight million and fifty, they may be trusted to supply a fine collection of blunders, many more than if the order were reversed and the larger number dictated first. The reason is clear. After the eight is fixed the pupil has in one case a clear field for disposing of the other figures, and in the other case a range which he has himself restricted.

It is the unusual that floors the pupils. Often a familiar fact seen from an unfamiliar point of view fails to be recognised, and evokes a wrong reaction. How else can we account for the fact that 35 percent of the children in Standard V who were set this simple test during the Board's recent inquiry gave a wrong answer?—"Add seventeen to seven thousand and write the answer in figures." A teacher is therefore well advised to take notes of the kind of mistakes which his pupils are liable to make, and to use these notes freely for testing purposes. The idea of the tests is not to inveigle the pupils into pitfalls but to rescue them from pitfalls. The pitfalls are there, anyway, and the pupils should be made acutely aware of their existence. Sometimes the only way to do this is to push the pupils in.

Let us now return to our system of place value as an expedient for representing a large number of numbers by a small number of symbols. Although the little children in the infant school and kindergarten learn something about place value, they cannot learn very much about it: its full significance is not borne in upon them till more mature years. All arithmetic begins with the natural number series, 1, 2, 3, 4, etc. There is no end to the series. There is an end to our capacity for naming them and for writing them down, but there is no end to the numbers themselves. And yet in a sense every number is simple, unitary, and unique. It is true that the name of the number indicated by 1765 is complex, and that the symbolism by which we represent it is complex: its name and symbol are manifestly compounded of simpler names and simpler symbols. But the number itself is a unique whole, and it is merely the need to economise thought that prevents us from giving it a single unique name. If I understand the Kantian philosophy aright (it would not be difficult to convince me that I don't), Kant in his *Critique of Pure Reason* supports his main argument by asserting that $5 + 7 = 12$ is not a statement of pure identity. It is not like saying that A is A, or that Jenkins is Jenkins: it is like saying that Jenkins is a genius. Kant didn't put it in that way. What he actually said was that $5 + 7 = 12$ is not an analytical but a synthetic proposition. He apparently meant that the two sides of the equation do not represent precisely the same idea. There is something on the right-hand side that is absent from the left-hand side. The 12 in fact is a new and unique number. But surely, you will say, 3 is simple in the sense that 12 is not. It is true that $12 = 3 + 3 + 3 + 3$; but then $3 = 1 + 1 + 1$. If nothing new is reached by adding, there is but one simple number, and that is the number 1.

Let us for a moment forget that numbers have either names or symbols, and connect the notion of numbers with pebbles, shells, or some other concrete and discrete objects, and let us try to imagine how our savage ancestors were compelled to count and to keep tally. Augustus De Morgan in the *Arithmetic* which he wrote a hundred years ago gives the following account of how our ten signs are made to represent other numbers besides the first nine:

"Suppose that you are going to count some large number, for example, to measure a number of yards of cloth. Opposite to yourself suppose a man to be placed, who keeps his eye upon you, and holds up a finger for every yard which he sees you measure. When ten yards have been measured he will have held up ten fingers, and will not be able to count any further unless he begin again, holding up one finger at the eleventh yard, two at the twelfth, and so on. But to know how many have been counted, you must know, not only how many fingers he holds up, but also how many times he has begun again. You may keep this in view by placing another man on the right of the former, who directs his eye towards his companion, and holds up one finger the moment he perceives him ready to begin again, that is, as soon as ten yards have been measured. Each finger of the first man stands only for one yard, but each finger of the second stands for as many as all the fingers of the first together, that is, for ten. In this way a hundred may be counted, because the first may now reckon his ten fingers once for each finger of the second man, that is, ten times in all, and ten tens is one hundred. Now place a third man at the right of the second, who shall hold up a finger whenever he perceives the second ready to begin again. One finger of the third man counts as many as all the ten fingers of the second, that is, counts one hundred. In this way we may proceed until the third has all his fingers extended, which will signify that ten hundred or one thousand have been counted. A fourth man would enable us to count as far as ten thousand, a fifth as far as one hundred thousand, a sixth as far as a million, and so on."

This clearly indicates one of the ways in which our system of notation may have come into being. It explains the principle of grouping and the principle of place value. The two do not necessarily go together. In our coinage and in our chaotic system of weights and measures we have one without the other: we have grouping without place value. Pence are grouped in twelves to form shillings, and shillings are grouped in twenties to form pounds. Inches are grouped in twelves to form feet, feet in threes to form yards, and yards in one-thousand-seven-hundred-and-sixties to form miles. Here we have grouping in plenty, but it is grouping of an arbitrary and lawless kind, very different from that in which our denary or decimal system is based. In that system 4625 expresses $4x^3 + 6x^2 + 2x + 5$ where x is 10; but 4 tons 6 cwts. 2 qrs. 4 lb. has no corresponding algebraic expression: it has no radix common to its terms, no law of regular progression from lower magnitude to higher. The consequence is that in our ordinary notation place or position fixes a scale of values, but in our weights and measures place tells us nothing. We have to fix labels to make up for it. Grouping is one thing, place value is another. The Romans had the one but not the other; and without the other little progress was possible.

A system having been invented by which place conferred upon a digit a value which was distinct from its name value, it became necessary to use some sort of scheme or framework into which we may fit a number so that the local value of each of its digits may become clear. In the Middle Ages an abacus was used which consisted of a board ruled into columns as in Figure 4. The I marks the units column, the X the tens, and the C the hundreds. I show in the figure how the following numbers would be indicated, 50; 3806; 609,017. Observe the triadic arrangement of hundreds,

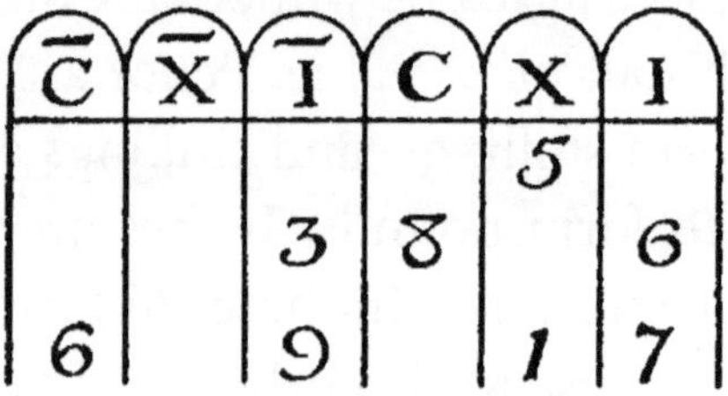

FIGURE 4. MEDIEVAL ABACUS.

tens, and units repeated on a higher level.

Another step was necessary to make the system complete. About the beginning of the thirteenth century the cipher (0) came into vogue, and that rendered the framework unnecessary. The cipher became a place-keeper, and by its use the place value of a digit could be indicated without the help of score or scaffolding.

In the following algebraic expressions $x = 10$. Pupils will find it a profitable exercise to write them as ordinary numbers:

$$2x + 4$$
$$5x^2 + 8x + 7$$
$$9x^3 + 4x^2 + 6x + 2$$

$$8x^4 + 9x^2 + x + 8$$

$$x^6 + 4x^4 + 5x^2 + 2x$$

$$8x^7 + 5x^3$$

$$4x + 3 + \frac{5}{x}$$

$$9x^2 + 8x + 3 + \frac{6}{x} + \frac{7}{x^2}$$

$$2x^3 + 3x + \frac{1}{x^3} + \frac{6}{x^4}$$

$$6x^5 + 3x^4 + x + 2x^{-1} + 3x^{-2}$$

The converse exercise of expressing algebraically such numbers as 37; 5060; 3.04; 18.0305, etc., helps to drive home the place-keeping function of zero.

The pupils should be brought to realise that it is the units figure that fixes the imaginary framework into which the digits are supposed to fit. They should get into the habit of regarding the units position (not the decimal point) as the base from which places are counted either to the left or to the right. Consider the following:

6	7	2	$\boxed{5}$	. 3	8	2	5
$6x^3$	$7x^2$	$2x^1$	$5x^0$	$3x^{-1}$	$8x^{-2}$	$2x^{-3}$	$5x^{-4}$

It is the number of removes from the units position that gives the index of x. The decimal point is not a "place" at all, but merely a device for marking the place of the units figure. So important is this units position that there is a growing tendency to emphasise it by putting 0 there even when it is not strictly necessary. Thus .036 is often written 0.036. So important is this position that it

69

gives to "standard form" its distinguishing mark and to logarithmic tables the key to the missing characteristics.

Between numerical notation and algebraic notation an interesting point of difference is observable. When symbols are placed simply side by side the missing sign in arithmetic is +, but in algebra it is ×. Thus 385¾ means 300 + 80 + 5 + ¾; but abc means $a \times b \times c$, and $(a + b)(c + d)$ means $(a + b) \times (c + d)$. In arithmetic it is addition that is taken for granted; in algebra, multiplication.

CHAPTER IX

SCAFFOLDING

But when he once attains the upmost round,
He then unto the ladder turns his back,
Looks in the clouds, scorning the base degrees
By which he did ascend.

SHAKESPEARE: *Julius Cæsar*.

A GOOD arithmetician solves his problems in the simplest and most direct way. He does without things which he once regarded as essential. He used to count on his fingers; he does so no longer. He used to mark his carried figures on his paper; he has given up this practice. He used to add by making up tens; he no longer needs ten as a resting-place in computation. Much that he worked in the margin of his paper he now works in his head. Things which were once a help first became superfluous, then became a hindrance. To get rid of them was an act of wisdom. Having no further use for the ladder by which he had climbed, he had kicked it away.

There are many such methods and devices in arithmetic—many little habits which begin by being friendly and end by being hostile. They helped to

71

build up a fabric of knowledge, but have ceased to be a part of the fabric. They are mere scaffolding. It is not always easy to say what is scaffolding and what fabric. Indeed, in an ultimate analysis, all our rules and methods would prove to be but props and planks round an ideal construction which is the real fabric. To a supreme intelligence all our complicated machinery of calculation—our symbolism, our notation, and our rules—would be as superfluous as is counting on the fingers to an ordinary man. Hence the real distinction is between scaffolding which may safely be pulled down, and scaffolding which cannot be pulled down without injury to the building.

Why should counting on the fingers be regarded with such disfavour? A child's fingers are his natural abacus—always at hand. That indeed is a great advantage, but it is an advantage which in time turns into a great disadvantage. The fact that fingers are always available, while matchsticks and counters are not, betrays the learner slowly and insidiously into a bad habit. Counting by units, whether the units are natural objects, or taps with the finger-tips, or accents in the mind, is excellent as an initial exercise, but execrable as a permanent practice. So fatally easy is it to count on one's fingers that many adults (and quite able adults, too) find it less trouble to add, say, 8 and 5 on their fingers than to reach 13 by a pure act of memory. The more the learner trusts to his fingers the less he trusts to his memory. In fact, he clings to a lower habit instead of climbing to a higher habit.

No habit in arithmetic is absolutely bad unless it

leads to a wrong answer. If it leads to the right results, it can only be relatively bad, in the sense that the route is larger and more difficult than it need be. The habit is bad because it prevents a better habit being formed. It is in this sense only that counting on the fingers is a bad habit.

Little children often shed their habits spontaneously: they give them up when they discover better ones. In counting objects there is a transitory stage through which all children seem to pass, a stage at which they have always to "start from scratch." When, for instance, a child is given a pile of four pennies and another pile of three pennies, and after he has counted each pile is asked: How many altogether? he does not take the pile of 4, which he has just counted, as the starting-point, and say 4, 5, 6, 7; he has to start at the beginning and count the four over again, saying, 1, 2, 3, 4, 5, 6, 7. Whether he suspects that the group of four pennies has changed since he counted it a few seconds ago I cannot say; whatever the reason, he counts it again. But no normal child stops at this stage. He finds his way out. He discards a bit of scaffolding.

Nearly all books on method advocate addition by making up tens. They say that $8 + 5$ should be worked thus: $8 + 5 = (8 + 2) + 3 = 10 + 3 = 13$. The *Handbook of Suggestions* puts it like this:

"If it is desired to add 'nine' and 'seven' together, the method will be somewhat as follows: 'One' is required to make 'nine' into 'ten,' and this 'one' taken from the 'seven' leaves 'six'; consequently, 'nine' and 'seven' are equivalent to 'ten' and 'six,' that is, to 'sixteen.'"[9]

This is a good device for beginners, just as useful as counting on the fingers, and just as dangerous. Yet none of the books in question strikes a note of warning; none of them points out that this trick of taking two bites at a cherry is a childish expedient which should be abandoned as soon as the habit of saying "nine and seven are sixteen" is firmly fixed. As an attempt to intellectualise the addition table the device in question is admirable; as a substitute for the addition table it cannot be too strongly condemned.

The intelligence that is supposed to be displayed by shifts to avoid memorising is illusory, or at best a short-range intelligence. Are we to assume that a child who arrives at "8 times 7 is 56" by saying that 4 times 7 is 28 and 8 times 7 is double 4 times 7 is necessarily more intelligent than the child who says straight off "8 times 7 is 56"? It is true that he is using his intelligence to make up for the defects of his memory; but it is not an intelligent use of intelligence to waste it on trifles. Why should we memorise the multiplication table and not the addition table? Apparently because the alternative to memorising the multiplication table is too ghastly to contemplate. If we were obliged to add eight sevens every time we needed the product 8 and 7, arithmetic would become intolerably tedious. The labour we save in memorising the addition table is not so obvious; but we save a little each time we add, and the cumulative savings are of no small consequence.

[9] *The Handbook of Suggestions*, p. 187.

The expediency of memorising any mathematical result, whether it be $6 + 7 = 13$, or $9 \times 12 = 108$, or $3^3 = 27$, or $(a + b)(a - b) = a^2 - b^2$, or $\log 3 = .4771$, or $\sqrt{2} = 1.414$, depends on the frequency with which we use it. The more frequently we need it the greater the advantage of carrying it about with us in our heads. It is the principle that underlies the fact that I always carry a penknife about with me, I sometimes carry a pair of scissors, and I never carry a corkscrew. The addition table is as useful as a penknife; so useful in fact that it should form part of the equipment of every cultured mind. What happens to most people is that they acquire it bit by bit by working sums. After having added 8 and 7 a large number of times in different contexts, and getting the answer by counting, or by noting that $8 + 7 = (2 \times 8) - 1$, or $= (2 \times 7) + 1$, or $= 10 + 5$, we can scarcely fail in the course of time to remember that the result is 15. What we need to do is to expedite the process—to get more quickly to the final stage, $8 + 7 = 15$.

When the analysis described in the above quotation from the *Suggestions* has been made once or twice it has served its purpose. The next step is to grasp the general principle that when 9 is added to another digit the units figure of the total is one less than the digit added. Then practice should take place with a row of separate figures like this:

4 2 8 9 7 3 6 0 5 1 4 7 9 6 8 3 5

Nine has to be added to each figure in turn, and the pupil says 13 11 17 18 16, etc. He must not say "9 and 4, 13"; he must not even think "9 and 4, 13." Knowing in the background of his mind that he has to add 9 to the 4, he simply thinks "13" as a single pulse of thought. The "9 and 4" is lumber which can be discarded, with profit to the free movement of the mind. Then the pupil should be practised in the repeated addition of 9. Beginning with 8, the series would be 17, 26, 35, 44, etc.

Similar exercises may be devised with 8 as the common addend.

It is sometimes recommended that in adding a column of figures we should search along the column for numbers which make ten. If there is a 2 near the bottom of the units column and an 8 near the top, these two should be coupled in the mind. So should any other pair, such as 7 and 3, or 4 and 6, that add up to ten. This is one of those "intelligent" methods which may be trusted to produce a fine crop of blunders. If the paired numbers happen to come together, that is a piece of luck that should not be overlooked; but to hunt for them is to hunt for trouble.

Didactic apparatus, discrete objects, diagrams, models, and illustrations are all in their times and their seasons valuable aids to efficiency; but if retained beyond a certain stage they seriously impede progress. They are little systems which, when they have had their day, should cease to be. The difficulty is to discern when they have had their day. The precise point at which they pass from help to hindrance differs with each individual mind; and it requires much insight and discrimination on the part of the teacher to discover the time when it is wise for the child to put away these childish things.

There are times when the mind of the pupil needs help and knows not where to seek it. There are times when a diagram would make a perplexing problem quite clear, but it never occurs to the pupil to make one. Even in mensuration (and a mensuration problem almost shouts for a diagram) the bulk of the pupils either work by rule of thumb or trust to their unaided imagination. In either case they are just as likely to get the answer wrong as right. If an average class of children about twelve years of age are given the length and breadth of a rectangle, a fair number of them will immediately multiply the two together, whatever the question may be. They will multiply them together just as cheerfully to get the perimeter as to get the area. A simple diagram would save them from that absurdity. Problems on the papering of rooms, the carpeting of floors, or the staining of borders may be made quite clear by drawing diagrams. Children who draw diagrams invariably get such sums right; but children who draw diagrams are amazingly few.

It is not only in mensuration that diagrams are useful: they clarify problems of all sorts. Look at this one:

Two quarts of milk are mixed with two quarts of water, and to half the mixture two more quarts of water are added. How much milk is there in a pint of the second mixture?

This problem, simple as it is, can be made clearer by drawing a diagram ABCD to represent the first mixture, and MNOP to represent half this mixture, the shaded part standing for milk and the unshaded for water. The completed figure MNQR, represents the second mixture, from which it is at once seen the ¼ of it is milk. The answer to the question is therefore "a quarter of a pint."

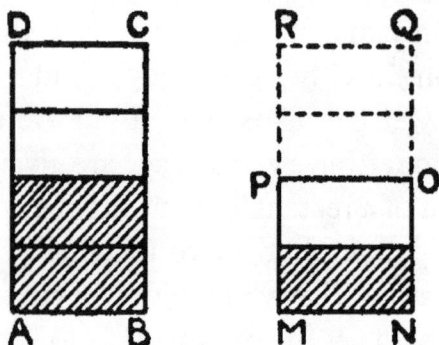

We see then that crutches are sometimes a bane and sometimes a boon. We see that once having used crutches in a mechanical way the pupil clings to them even when he can walk better without them; but when he needs other crutches for a long and difficult leap, he neglects to use them. Both the use and the neglect are due to mental inertia—one to the inertia of motion, the other to the inertia of rest. The mind is in the first case too lazy to change an old habit; in the second too lazy to initiate a new one.

CHAPTER X

SUBTRACTION

A woodpecker, who had bored a multitude of holes in the body of a dead tree, was asked by a robin to explain their purpose.

"As yet in the infancy of science," replied the woodpecker, "I am quite unable to do so. Some naturalists affirm that I hide acorns in these pits; others maintain that I get worms out of them. I endeavoured for some time to reconcile the two theories; but the worms ate my acorns, and then would not come out. Since then I have left science to work out its own problems while I work out the holes."

AMBROSE BIERCE: *Fables.*

I understood subtraction best, and for this I had a very practical rule—"Four from three won't go, I must borrow one"—but I advise everyone, in such a case, to borrow a few extra shillings, for one never knows.

HEINE: *Reisebilder.*

What do we do at school? We add and we un-add.

EVE McCOWEN, *aged six.*

TIME was when simple subtraction gave the teacher little trouble. The children were told to borrow ten and pay one back, and they cheerfully obeyed. They never questioned the logic of the matter. They were assured that that was the way to do it; they discovered that it gave the right answer; and that was enough for them. And they got their subtraction sums right. But to-day a large number of quite old children can't get them right. Hear the evidence of the Board's Inspectors.[10] This sum was set to a large number of children in Standards V and VII: "Take thirty-four thousand and eight hundred and forty-six from fifty-seven thousand eight hundred and forty-nine." Here we have a simple straightforward sum in subtraction; and yet only 66 percent of the children in Standard V got it right, and only 83 percent of those in Standard VII. I can recall the time when Standard II could do better than that. The accomplishments of Standard II were lamentably few, but they included a mastery of simple subtraction.

Subtraction is weak; that is certain. It is also certain that the root of the weakness lies in the method employed. Fifteen years ago two-thirds of the schools in London used a confusing and cumbersome method known as "decomposition." To-day the method is rapidly passing into oblivion—except in the infant schools. Taking the whole of the kingdom, it is probable that about half the schools still use this method. I have watched children near the top of the senior school subtracting, say, 17 from 30,000 and heard them mutter: "7 from 0, cannot, go next door. Nothing there; go next door. Nothing there; go next door." And they continued this Mother Hubbard business till they got to the 3. Then they put a little 2 over the 3, 9 over each of the next three noughts, and 10 over the last nought. And this is regarded as a

[10] *The Teaching of Arithmetic in Elementary Schools,* p. 15.

more intelligent procedure than borrowing ten and paying one back. The truth is that the guiding formula is as meaningful, or as meaningless, in one case as in the other.

The main charge against the decomposition method is that it doesn't work. It fails to give a reasonable measure of facility and accuracy. I have elsewhere [11] stated the grounds on which I make this assertion, and it will suffice if I give here a brief account of the evidence, adding thereto the evidence that has subsequently come in.

Just before the War I carried out an extensive research into the ability of London children to perform the fundamental processes of adding, subtracting, multiplying, and dividing. The tests consisted of simple mechanical sums to be worked on paper. When the results were examined, it was found that the schools fell into two distinct groups—those in which subtraction was good, and those in which it was bad. By "good" or "bad" I mean good or bad in comparison with the achievements of the same schools in the other three processes. On fully investigating the matter I found that where subtraction was good it was taught by the old-fashioned method of equal additions, and where it was bad by the new-fashioned method of decomposition. It will be seen from the accompanying graph that the difference between the two types of schools is considerable. The score on the vertical ordinate gives the number of simple operations (such as $8 - 3 = 5$) rightly performed in five minutes. The lower graph

[11] *Mental Tests*, pp. 168-174.

81

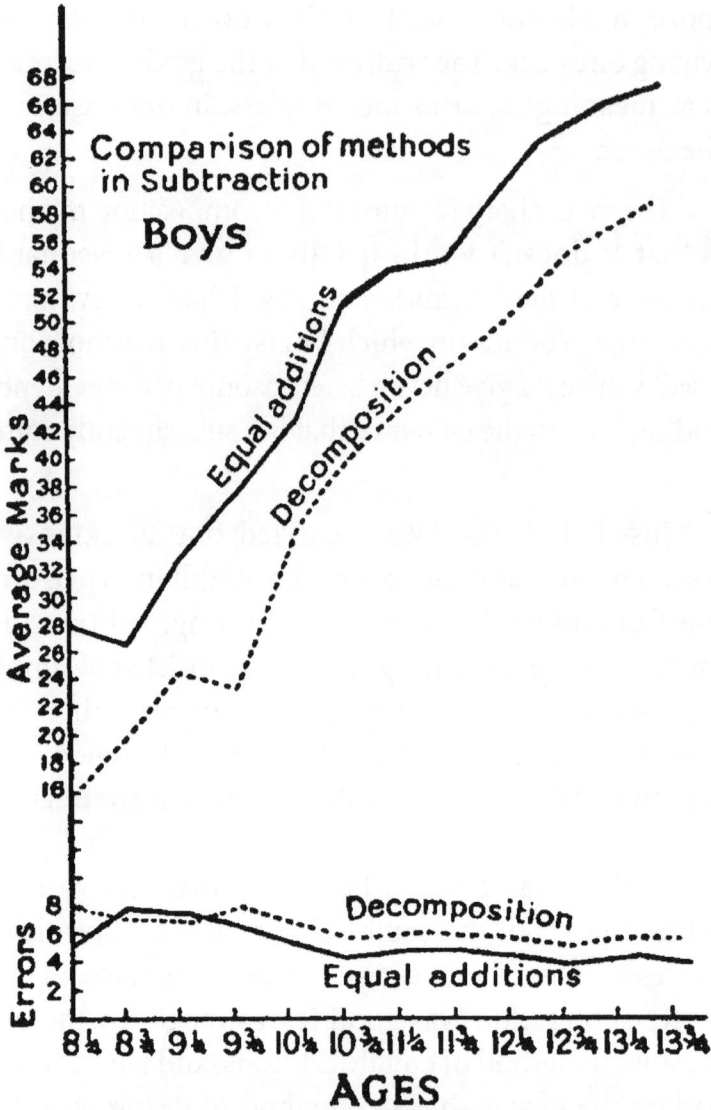

Comparison of methods in Subtraction

Boys

Equal additions

Decomposition

Average Marks

Decomposition

Equal additions

Errors

AGES

records the number of errors. It is clear that both in point of accuracy and in point of speed the advantage lies on the side of equal additions. At the early ages it is at least 50 percent more efficient.

82

Five years later I repeated the experiment in another London district with a revised edition of the tests and found the same remarkable disparity between the results of the two methods. It was possible by merely glancing at the tabulated results to identify the schools in which the method of decomposition was taught. There was always the tell-tale "slump" in subtraction.

Later on the same tests were again set by other people, to other children, in another part of England. In 1925 they were administered to 118 elementary schools in the County of Gloucestershire by His Majesty's Inspector and his staff.[12] Broadly speaking, the results were the same as in London. And there was the same sheep and goat division of the schools. The verdict of Gloucestershire on the question of method in subtraction may be inferred from the following table:

	Decomposition.	Equal Additions.
% schools above average	27	51
% above 2 sums in 3 minutes	51	73
% of very bad results, i.e. under 1.5	27	9

The averages obtained by the two types of methods were:

	Schools.	Average in 3 minutes.
Decomposition	27	51
Equal additions	51	73

[12] See *Method and Results in Arithmetic,* by H. J. Larcombe, pp. 93-94. Mr. Larcombe was one of the inspectors engaged in the investigation.

It was a pure coincidence that half the schools examined had adopted one method and the other half the other: the schools were not chosen on a basis of method.

So far the evidence has been secured by extensive research. I will now present evidence which was secured by intensive research—evidence which is not only confirmatory but, I venture to think, conclusive. At the time when I was making my first investigations, Mr. W. H. Winch was quite independently investigating the same problem by the mode of research which he had himself invented—the method of "equal groups." This particular technique affords the most precise means of scientific research in education that has yet been devised. A full account of Mr. Winch's investigation appears in the issues of the *Journal of Experimental Pedagogy* for June 1920 and for December 1920. I would strongly advise those who have any lingering doubts on the question to read these two articles.

Mr. Winch first experimented with two classes of older girls (girls in Standards V and VI*b*) who knew one method only—the method of decomposition. He classified them in two equal and parallel groups on the basis of their proficiency in subtraction. One of these groups was given eight short lessons in equal additions, and the other the same number of lessons in decomposition. At the close the equal-additions section was ahead of the other. Here are Mr. Winch's conclusions:

"(1) The method of equal additions in subtraction taught to children late in school life, who have hitherto worked by decomposition, produces results in a few weeks equal, on the whole, and superior in the weaker children, to those produced by the method of decomposition.

"(2) The amount of the gain involved does not justify a change of method at this late period of a child's school career."

The second experiment was of greater importance, for it was made on younger children, children whose average age was eight and a half, and many of whom had quite recently come up from the infant school, where the method of decomposition was taught. The experiment proceeded on the same lines as before, and at the end of eight lessons it was discovered that the equal-additions group was 30 percent in advance of the others. Mr. Winch sums up his conclusions thus:

"The method of equal additions shows to decided advantage with young children in accuracy and rapidity; and this is true both in the case of the superior children, who already had learnt something of both methods, and also in the case of the inferior children, who, prior to the experiment, really knew nothing of either method."

In view of these facts how are we to account for the vitality of the method of decomposition? The answer is that it is the easier method to understand and the easier method to begin with. It was born in the infant school, and it flourishes in the infant school; and the senior school adopts the method, not through any fondness for the method itself, but from sheer reluctance to change.

Let us consider the simplest form of subtraction "with borrowing." Take, for example, $\frac{34}{18}$. Children in the infant school or kindergarten are nowadays accustomed to represent the 3 by three bundles of sticks with ten in each, or by strings of beads, or by some other concrete material; and in order to subtract the 8 they are taught to untie one of the bundles and to regard thirty-four as twenty-fourteen. Subtraction then becomes easy. 8 from 14 leaves 6, 1 from 2 leaves 1.

So far all is plain sailing. Later on, however, when the children reach the senior school, they will have to deal with hundreds, and the minuend might be 304 instead of 34. If 18 has to be subtracted as before, a large bundle of 100 has now to be decomposed into tens, and one of the tens decomposed into units before subtraction can begin at all. When the minuend is 30,004 the amount of decomposition that has to take place makes one wonder how the sum can survive it. As a matter of fact it rarely keeps itself together without being propped and buttressed by a series of supporting figures.

The ineffectiveness of the method of decomposition is not, however, solely due to the amount and complexity of the changes that have to be made in the minuend: there are other causes which can best be appreciated by contrasting the two methods. In both methods the procedure rests in a principle of compensation. We alter the top figure, and we have to compensate for the alteration. In the first example given above (34–18) we have to change the 4 into 14 whatever method we adopt, but we compensate in one method by making the 3 one less, and in the other by making the 1 one more. And it seems at first blush as though one mode of compensation were just as easy as the other. As a matter of fact, however, the equal-additions compensation is the easier, first because it is easier to add 1 than to subtract 1, and secondly because it has a slight advantage in point of time. The compensation is made a little earlier. When there is a series of noughts in the minuend the advantage is by no means slight: the compensation is made not a little earlier but very much earlier. Accounts are settled with much greater dispatch. One is "cash payment," the other "credit." And the postponement of the compensating act increases the chances of its fulfilment being forgotten.

We may well, therefore, leave decomposition to die in peace (though it seems rather slow in doing it) and turn our attention to teaching the method of equal additions. Some infant-school teachers assure me that to teach it as a reasonable process is a difficult task; others assure me that it is quite easy. It is evidently not impossible. What we first have to do is to inculcate the general principle that the difference between two numbers remains unchanged if the same number be added to each. This may be done in a variety of ways. It may, for instance, be pointed out that if there are two brothers who are five and eight years of age respectively there will always be three years' difference between them, however many years pass over them.

THE ESSENTIALS OF ARITHMETIC

$8 - 5 = (8 + 1) - (5 + 1) = (8 + 2) - (5 + 2)$, etc.

Taking our stock example $\frac{34}{18}$, we explain our practical

procedure of taking 8 from 14 and then taking 2 from 3 by stating that we add ten to each line—in the form of 10 units to the top line, and in the form of 1 ten to the bottom.

There is another way of explaining it. As I have lent 10 to the top line, I must take it away again at the first opportunity, that is, when I take the 1 away.

Finally, there is the explanation which is to young children the most satisfying of all. It silences all inquiry and dispels all doubts. It is this: I *give* ten to the top line (they like generosity), and *to be fair* (they dislike favouritism) I give ten to the lower line as well. This satisfies not only their reason, but something which is very much stronger—their sense of justice. When it happens to be money, they will possibly regard it as "putting a bob each way," if I understand the phrase aright (which I probably do not).

When the first step is understood, the second step is quite easy to understand. The reasoning is the same; we simply add 100 to each line instead of adding 10.

The probability is that some children will understand the reasoning, others will not. It doesn't much matter. They will forget it in any case. Forget it, I mean, when actually working sums. They might remember it if challenged, or pressed for a reason, but for all practical purposes they will simply have at the back of their minds a vague conviction that what they are doing has the sanction of reason and will proceed to work by rule of thumb. Just as you and I do. Even granting that the decomposition method is easier to understand, it by no means follows that it is the better to use. It is easier to understand a boy's drum-and-string telephone than the ordinary telephone; it is easier to understand a Stephenson's Rocket than a Rolls-Royce car; it is easier to understand a sailing-vessel than a turbine steamer.

The really important thing we have to teach is not equal additions as a logical process, but equal additions as a mechanical habit. We can get along better without the former than without the latter. It is always a wise plan to divide difficulties into as many parts as possible and to tackle each part separately. The first difficulty the pupil has to meet is a very little one indeed: he has to be able to recognise at a glance when the figure in the minuend has to be increased by ten. He should not have to say (taking our stock example) "8 from 4, I cannot, add 10." He should not even have to say it mentally. He should simply glance at the figures and say "8 from 14." Put on the blackboard a tremendous subtraction sum like this (it's only for practice):

$$8 \quad 3 \quad 2 \quad 5 \quad 6 \quad 4 \quad 7 \quad 0 \quad 3 \quad 1 \quad 3 \quad 5 \quad 6 \quad 4 \quad 8 \quad 7$$
$$2 \quad 7 \quad 6 \quad 8 \quad 9 \quad 5 \quad 8 \quad 1 \quad 4 \quad 2 \quad 5 \quad 6 \quad 7 \quad 6 \quad 9 \quad 8$$

Except in the highest denomination each figure in the subtrahend is larger than the corresponding figure in the minuend. Therefore the top figure has in each case except the last to be increased by ten. The pupils should be required to make the increase automatically by saying in succession: seventeen, eighteen, fourteen, sixteen, etc.

They should not say 10 and 7, 17, but simply 17. They should then be practised in automatically increasing the lower figure by one without going through the stage of saying 9 and 1, 10; 6 and 1, 7; etc. They should be pressed to say rapidly 10, 7, 8, 7, 6, 3, etc. The next stage is to practise both operations together, thus: 8 from 17, 10 from 18, 7 from 14, 8 from 16, 7 from 15, etc. There is no need to carry out the actual subtraction here. We are engaged in fixing a habit, not in working a sum. The next practice should be with another long example in which "borrowing" is sometimes necessary and sometimes not. Here is one:

9 6 0 3 8 0 5 2 7 5 3 3 5 1 4 8
3 7 1 2 9 8 0 7 0 8 2 8 6 0 9 4

This is an exercise in change and compensation for the change, not in subtraction. The pupil should merely say: 4 from 8, 9 from 14, 1 from 1, 6 from 15, 9 from 13, etc.

I do not suggest that the pupils should become expert in such exercises as I have given above before they are allowed to work simple sums in subtraction. The two should proceed side by side. The exercises are proposed not so much to make subtraction possible as to make it automatic—to facilitate the passage from the stage when it is difficult to get the sum right to the stage when it is difficult to get it wrong.

The following extract is taken from the Appendix to Augustus De Morgan's *Elements of Arithmetic*:

"SUBTRACTION. The following process is enough. The carriages, being always of one, need not be mentioned.

From 79436258190	8 and 2′, 4 and 5′, 7 and 4′,
Take 58645962738	3 and 5′, 6 and 9′, 10 and 2′,
20790295452	6 and 0′, 4 and 9′, 7 and 7′,
	9 and 0′, 5 and 2′. It is useless

to stop and say, 8 and 2 make 10; for as soon as the 2 is obtained, there is no occasion to remember what it came from."

The usual way is this: 8 from 10, 2; 4 from 9 5; 7 from 11, 4; etc. De Morgan's method is an adding method, and is probably the most efficient method of all. At any rate it is the method usually adopted when much computation has to be done. Its proper name is the Method of Complementary Addition; but I hesitate to call it that because the term seems to be nowadays applied to another method which is neither adding nor subtracting, but a mixture of both. By this method the above example would be worked thus: 8 from 10, 2; 4 from 9, 5; 7 from 10, 3, and 1, 4; 3 from 8, 5; 6 from 10, 4, and 5, 9; 10 from 10, 0, and 2, 2; etc. In other words, we subtract from no number higher than 10. The advantage seems to be that there is no need to burden the memory with subtraction tables higher than 10. But this advantage is counterbalanced by serious disadvantages. It is long and cumbersome; two steps are taken when only one is necessary; and the relief to the memory is insignificant, unless we curtail the addition table in like fashion, and that few would agree to. If, therefore, we have to learn $8 + 5 = 13$, it suffices, on De Morgan's plan, for both addition and subtraction. And on the ordinary plan the step from $8 + 5 = 13$ to $13 - 8 = 5$, or $13 - 5 = 8$, is not difficult to achieve. In fine, there is no real economy. What is saved in memory is wasted in ceremony.

There is, however, much to be said for this mixed method when we have to subtract money or weights and measures. If, for instance, we have to take £6 15s. 9d. from £48 13s. 5d., it is quite a good plan to say: 9d. from 1s. leaves 3d., and 5d., 8d.; 16s. from £1 leaves 4s., and 13s., 17s.

To the question, Which is the best method for the infant school or kindergarten? one hesitates to give a dogmatic answer. The modern infant school lays so sound a foundation for future work that it would be ungracious to carp at the teachers choosing what seems to them to be the more rational way. Taking the long and generous view, they would be well advised to teach the method of equal additions (and they are in fact teaching this method in increasingly large numbers); but even if they don't, it is no hard task to change the method when the senior school is reached. As a mode of procedure it simply means increasing the bottom figure instead of decreasing the top figure. A child can readily be brought to see that it makes no difference to the answer whether we make the bottom figure one more or the top figure one less. And while we can always make the bottom figure one more, we cannot always make the top figure one less. We cannot when it is nought.

CHAPTER XI

THE MULTIPLICATION TABLE

Moth. How many is one thrice told?
Armado. I am ill at reckoning; it fitteth the spirit of a tapster.
SHAKESPEARE: *Love's Labour's Lost.*

The most Devilish thing is 8 times 8, and 7 times 7 is what
nature itself can't endure.
MARJORY FLEMING'S DIARY
in JOHN BROWN'S *Horæ Subsecivæ.*

To memorise the multiplication table without understanding it is worse—much worse—than understanding
it without memorising it. Both happen; the first rarely,
the second frequently. To know the tables parrot-fashion
is not arithmetic at all: it is as yet a mere possibility
of help—or of hindrance. A child whose mind is
cumbered with undigested tables, with bonds which
he has not himself bound, when asked how many 3 and
5 make is just as likely to say 15 as 8. Fortunately, the
children in the modern infant-school or kindergarten
receive so excellent a grounding—are brought so
closely and so actively into contact with reality—that
operations in arithmetic cannot fail to have for them a

real meaning. They build up the multiplication tables so that a pair of factors such as 4 × 8 becomes a challenge to an experiment with beads or beans or match-sticks. They then get a clear insight into the structure of the multiplication tables. Each bond becomes a record of an adventure which they can repeat with a feeling of certainty that they will always arrive at the same goal.

It is quite easy to discover when a child is able to interpret the tables aright. Ask him to construct an entirely new table, such as the thirteen times. If he attacks the task in the right manner straight off, it is time for him to take the next step. And the next step is to get (if he has not got it already) a good grip of the commutative law: $a \times b = b \times a$. It can be arrived at by reading an arrangement of objects in two ways. The asterisks in Figure 5 may be regarded as three rows of four or as four columns of three. In either case the

```
    *       *       *       *

    *       *       *       *

    *       *       *       *
```

FIGURE 5

total is 12. That is 3 × 4 = 4 × 3. His comprehension of this law may be tested by such oral questions as these:

If 14 × 17 = 238, how much is 17 × 14?

If 25 × 362 = 9050, how much is 362 × 25?

Most teachers pass abruptly from this stage into the memorising stage. Now that the children know what the tables mean, the next step should be to learn them by heart. But I think we should wedge in a little more rationalising between the two stages. Though I strongly hold that the tables should be made mechanical, I don't think they should be mechanically made mechanical: I repudiate the view that a thoughtless repetition is the best way to secure automatism. What we aim at is a strengthening of the bridge made in the mind between, say, 9 × 7 on the one side and 63 on the other, and the mere adding together of nine sevens or of seven nines and finding that they come to 63, though necessary, is not enough. The bond is made, but not secured. It needs strengthening; and it is strengthened by noting certain peculiarities of all multiples of nine, and by visualising the place of 9 × 7 on the number chart. Any rational connection that is grasped by the pupil becomes in itself the best sort of connective tissue. The more thought we can get the pupils to put into the bridge-building at first, the more serviceable will the bridge be at the finish. Though the bridge has to be used without thought, it does not mean that it has to be built without thought. Though the bond is unreasoned, it should none the less be felt to be reasonable. Repetition is necessary, too, of course, but repetition is much more effective when it is thoughtful and purposeful than when it is a careless reiteration of sounds without sense.

I suggest, therefore, a transition stage where one of the aims should be to give the child a rough idea of the magnitude of a product by visualising the place of that product in the numerical system. Miss Punnett has an excellent way of presenting the system to the eye. As a frontispiece to her book on Arithmetic she gives a number chart where a hundred coloured discs are arranged in rows of ten, all in the same row being of the same colour; and she recommends that children should be practised in rapidly marking off a given number, such as 12, 26, 33, 67, etc.[13] In this chapter the earlier items of the natural number series are presented in similar charts, with numbers substituted for the coloured discs. These charts are intended to be used for teaching tables.

The best order in which to build up the tables is not the best for teaching from the chart, nor yet the best for memorising by repetition. When building up it is easiest to begin with the "twice" and follow up with the 3 times. In using the charts, however, and in learning by heart, the "10 times" affords the best starting-point. The pupil should first count in tens down the chart (see Chart I), 10, 20, 30, etc. He should then say: one ten is 10, two tens are 20, three tens are 30, etc. Then in this

1	2	3	4	5	6	7	8	9	10
11	12	13	14	15	16	17	18	19	20
21	22	23	24	25	26	27	28	29	30
31	32	33	34	35	36	37	38	39	40
41	42	43	44	45	46	47	48	49	50
51	52	53	54	55	56	57	58	59	60
61	62	63	64	65	66	67	68	69	70
71	72	73	74	75	76	77	78	79	80
81	82	83	84	85	86	87	88	89	90
91	92	93	94	95	96	97	98	99	100

CHART I

[13] *The Groundwork of Arithmetic*, pp. 81-86.

form: 10 ones are 10, 10 twos are 20, 10 threes are 30, etc. An average child would by this time know the 10 times, and even know it "didgy-dodgy," as the children say. That is, he would be able to reply to: 7 times 10? without having to start from the beginning. The 5 times should come next (Chart I), and after that the twice, 4 times, and 8 times tables (Chart II).

1	2	3	4	5	6	7	8	9	10
11	12	13	14	15	16	17	18	19	20
21	22	23	24	25	26	27	28	29	30
31	32	33	34	35	36	37	38	39	40
41	42	43	44	45	46	47	48	49	50
51	52	53	54	55	56	57	58	59	60
61	62	63	64	65	66	67	68	69	70
71	72	73	74	75	76	77	78	79	80
81	82	83	84	85	86	87	88	89	90
91	92	93	94	95	96	97	98	99	100

CHART II

Then come the other related tables, the 3 times, 6 times, and 9 times. (See Chart III.)

It will be observed that the tables form patterns on the chart. The 9 times can be traced by a diagonal line, which is crossed at right angles by the 11 times line. The 11 times table, though interesting, is of little practical importance. The 12 times table, however, is of cardinal importance, as we have traces of a duodecimal system in our dozen and our gross, and in the number of pence in a shilling and of inches in a foot. The 12 times forms

1	2	3	4	5	6	7	8	9	10
11	12	13	14	15	16	17	18	19	20
21	22	23	24	25	26	27	28	29	30
31	32	33	34	35	36	37	38	39	40
41	42	43	44	45	46	47	48	49	50
51	52	53	54	55	56	57	58	59	60
61	62	63	64	65	66	67	68	69	70
71	72	73	74	75	76	77	78	79	80
81	82	83	84	85	86	87	88	89	90
91	92	93	94	95	96	97	98	99	100

CHART III

a pattern similar to the 8 times. It is a profitable exercise to trace these two patterns on a fresh chart.

The relation between the various tables may be exhibited in another way. The alternate products in the twice table form the consecutive products in the 4 times table, and the alternate products in the 4 times the consecutive products in the 8 times. (See Chart IV.)

The 3 times, 6 times, and 12 times may be shown to be connected in a similar way.

There are certain peculiarities of the 9 times table which well repay study. The sum of the digits in the product is always 9, except in $11 \times 9 = 99$, where it is double nine. If we draw a line between 9×5 and 9×6, we divide the table into symmetrical halves with the digits below the line reversing the digits above the line. Thus 45 is balanced by 54, 36 by 63, 27 by 72, 18 by 81, and 09 by 90. The product of 9 and n, when n is any number from 2 to 10 inclusive, consists of two figures,

$2 \times 1 = 2$
$2 \times 2 = 4$. $4 \times 1 = 4$
$2 \times 3 = 6$
$2 \times 4 = 8$. $4 \times 2 = 8$. $8 \times 1 = 8$
$2 \times 5 = 10$
$2 \times 6 = 12$. $4 \times 3 = 12$
$2 \times 7 = 14$
$2 \times 8 = 16$. $4 \times 4 = 16$. $8 \times 2 = 16$
$2 \times 9 = 18$
$2 \times 10 = 20$. $4 \times 5 = 20$
$2 \times 11 = 22$
$2 \times 12 = 24$. $4 \times 6 = 24$. $8 \times 3 = 24$

CHART IV

the first of which is $n - 1$ and the other $9 - (n - 1)$. For example, 9 eights are seventy-something, and that something is $9 - 7$. It is good practice for a pupil to construct the whole table from this formula.

The 3 times table has some of the properties of the 9 times. The sum of the digits in the product is always divisible by 3. The sums form a rhythm down the scale, thus: 3, 6, 9; 3, 6, 9; 3, 6, 9; 3, 6, 9.

The 7 times table is, as Marjorie Fleming had discovered, the most intractable of the lot. The pattern it forms on the number chart has about as much regularity as the knight's moves on a chessboard. Nor is there any simple test of divisibility by seven as there is of the other digits.

The tables might sometimes be presented, or written, for variety's sake, in the form given in Chart V.

3 *times*	6 *times*	9 *times*
1 2 3	1 2 3 4 5 6	1 2 3 4 5 6 7 8 9
4 5 6	7 8 9 10 11 12	10 11 12 13 14 15 16 17 18
7 8 9	13 14 15 16 17 18	19 20 21 22 23 24 25 26 27
10 11 12	19 20 21 22 23 24	28 29 30 31 32 33 34 35 36
etc.	etc.	etc.

CHART V

All these modes of presentation and practice are not merely aids to understanding: they are aids to memory also. But aids only. The tables have to be memorised by brute repetition, accompanied by a desire and an effort to remember. Such memorising is not easy. If anyone thinks so, let him try to learn by rote an entirely new table, such as the 19 times. The dullness of the task and the small meed of success will arouse a sympathy with young children in their efforts to remember. Mr. Bertrand Russell confesses that as a child he wept bitterly because he could not learn the multiplication table. He is one of a huge army. Hence the need for bringing in that spirit of sport to which I have referred in a previous chapter. Tackled in the right frame of mind, and in proper quantities, and perhaps without too much insistence on attentive effort, the tables are by no means a bugbear. As Professor Nunn assures us, "the young teacher may safely disregard the view that the repetition of tables, dates, grammatical paradigms, arithmetical or algebraic operations is unpedagogical because it has to be forced upon unwilling nature. The child who rejoices in his power to repeat the jingle Ena, dena, diva, do will not fail to delight in a mastery over more serious forms of routine." [14]

In considering the question, Which is the best way for children to memorise the tables? we must not forget that each item is a little system in itself, which should be learnt independently of all the other little systems. The important thing about 7 times 9 is that it equals 63, not that it comes after 7 times 8 and before 7 times 10. It follows that the order in which the items are repeated should be varied. They should be repeated backwards and forwards and "didgy-dodgy."

I have discussed in a previous book[15] certain methods of learning the tables and have given reasons for preferring individual methods to collective methods. There is no doubt something to be gained by simultaneous repetition; but there is much more to be gained by individual learning and individual testing. It is important that records should be kept of each child's proficiency in the tables. The child should know precisely how much of the tables he knows thoroughly, how much he knows imperfectly, and how much he doesn't know at all. The more he takes an interest in his own progress the more that progress is accelerated.

Mr. Winch has demonstrated that a good way to learn the tables is to apply them at high speed. If the 6 times table is placed before a child who doesn't know the 6 times table, and he is put to work a large number of examples of the type $42,805 \times 6$, and to work against time, he will find himself memorising the table as a

[14] *Education: its Data and First Principles,* pp. 62-63.

[15] *Mental Tests,* pp. 176-178.

saving of time and a defence against drudgery.

Some years ago I tried to discover which, among all the bonds in the multiplication table, is the most difficult to remember. $2 \times 1 = 2$ seemed the easiest, and $2 \times 2 = 4$ not much harder, but which was the hardest? My method of inquiry was to dictate all the items in mixed order and at a fixed rate, and then note the incidence of error. The presumption was that the item most frequently wrong was the item most difficult to remember. I expected 8×7 or 9×8 or some such combination of high digits to come out on top as the most difficult. But I was disappointed. Nothing invariably, or even generally, came on top. The order was different in every class I tested. I came to the conclusion that I was not measuring inherent difficulty at all, but rather certain accidents of teaching or of opportunity. Inherent difficulty was one of the factors, but its measure was falsified by variations in the other factors.

These conclusions are confirmed by investigations carried out on a much larger scale in America. Professor Clapp, of Wisconsin University, tested over 6,000 children in the way I have described above; but with this difference: he included noughts.[16] He thus used 100 combinations altogether, ranging from 0×0 to 9×9. He found, as I found, that the results varied from grade to grade. Massing the whole returns, he found the most difficult bond to be 7×0. After that came in descending

[16] *The Number Combinations: their Relative Difficulty and the Frequency of their Appearances in Textbooks.* By Frank L. Clapp. (University of Wisconsin.)

order of difficulty 0×5, 0×7, 0×1, 4×0, and 0×8. The first bond that did not contain nought was 9×7, the 17th in order of difficulty, and the next 8×7, the 20th. All these were time tests: the items were dictated at a given rate and no time was allowed for deliberation. Professor Clapp now proceeded to test the strength of the bonds in another way. He let the examinees work multiplication sums at their own rate and in their own way, and then counted the errors. By this means he got a new list, which differed considerably from the first. The list was headed this time by 0×2, and the next five were 9×4, 9×7, 8×6, 8×8, and 7×6. The first on the first list (7×0) came 42nd on the second list. And the bonds with noughts did not appear anything like as formidable in the second investigation as in the first.

Professor Clapp's final step was to analyse the textbooks used by the children he had tested, and to compare the frequency with which the bonds appeared in the textbooks with the frequency with which errors appeared in the tests. And he found a negative correlation. He found that, according to his criticism, the textbooks gave abundant practice in the easy bonds and scanty practice in the difficult bonds. And he blamed the textbooks. It did not seem to occur to him that there was another possible explanation; that the textbooks themselves were responsible for what he discovered; that he was not really finding the relative difficulty of the bonds, but the relative amount of repetitive strengthening the bonds had received.

A still more interesting piece of research has been carried out by Batson and Combellick at the University of South Dakota.[17] Here the subjects of the experiment were not children, but 83 graduates and undergraduates at a university—adults whose habits of multiplying numbers had already been pretty well fixed. They had to reply to 50,000 questions altogether—questions of the form, $5 \times 3 = ?$ The object was, not to discover whether they knew the tables or not, but to measure the speed with which they responded. The assumption was that the greater the hesitancy the greater the difficulty of the combination. It was found that on the average it took 3.11 tenths of a second to respond to 1×3, and 6.11 tenths of a second to respond to 6×9. It was assumed, therefore, that 6×9 is twice as hard to remember as 1×3. As might be expected, errors were infrequent; but such as they were about a quarter of the whole lot occurred in combinations containing zero. In other words, the most awkward number to occur in a bond is nought. The other figures follow in this order: 9, 7, 8, 4, 6, 5, 1, 3, 2. Apparently 1 is not the easiest digit to deal with. Nor is 6 as hard as 4. Leaving errors out of account and taking time of response as the sole criterion, the six most difficult items in the multiplication table are: 6×9, 4×4, 9×0, 0×6, 9×7, and 8×7.

These researches, dull and inconclusive as they are, force upon our notice one solid fact, the fact that the table which needs most attention from us at the present time is not the nine times, nor even the seven times, but the nought times. The nought times is the table that most frequently breaks down in practice, and yet we consider it so easy that we never teach it at all. Even the

---

[17] *Journal of Educational Psychology,* vol. xvi, No. 7, pp. 467-481.

once table is not so easy as it seems: it too gives ample occasion for stumbling. Hence the table-book of the future is not at all unlikely to begin as in Chart VI:

$$0 \times 0 = 0 \quad 1 \times 0 = 0 \quad 2 \times 0 = 0 \quad 3 \times 0 = 0$$
$$0 \times 1 = 0 \quad 1 \times 1 = 1 \quad 2 \times 1 = 2 \quad 3 \times 1 = 3$$
$$0 \times 2 = 0 \quad 1 \times 2 = 2 \quad 2 \times 2 = 4 \quad 3 \times 2 = 6$$
$$0 \times 3 = 0 \quad 1 \times 3 = 3 \quad 2 \times 3 = 6 \quad 3 \times 3 = 9$$

etc.   etc.   etc.   etc.

CHART VI

Another important fact emerges from the inquiry: it is that the commutative law does not work automatically. The fact that a child knows four times seven as a fixed mechanical habit is no guarantee that he knows seven times four as a fixed mechanical habit. Nor is the strength of one of these bonds any measure of the strength of the other. They have to be learnt as separate items and tested as separate items. Though they have to be understood as related, and though this understanding is in itself a vehicle of transfer from one to the other, yet it is found necessary to fix each by repetition as though it were an independent system.

Among the many hints and suggestions contained in this chapter the most important is: Don't forget the noughts.

CHAPTER XII

MULTIPLICATION

The old order changeth, yielding place to new.

TENNYSON: *Morte d'Arthur.*

As a general rule, a traditional method rests on a solid basis of experience. It is the method that has survived. Having killed off its competitors, it holds the field alone. And yet it is never safe to assume that the struggle is over, to assume that no new competitor will arise more formidable than the rest and drive the champion off the field. This seldom happens in the rudiments of arithmetic; but it does sometimes. It is happening now in simple multiplication. There is at present a struggle going on in the schools between two rival methods of multiplying by the higher numbers—numbers beyond nine. The traditional method is to begin with the unit figure of the multiplier: the newer method is to begin at the other end. Let me illustrate by an example where the digits of the multiplier are made the same in order to supply a simpler comparison of values.

At first blush there does not seem to be much

Old Method	New Method
3 5 9 7	3 5 9 7
6 6 6	6 6 6
2 1 5 8 2	2 1 5 8 2
2 1 5 8 2	2 1 5 8 2
2 1 5 8 2	2 1 5 8 2
2 3 9 5 6 0 2	2 3 9 5 6 0 2

difference between the two. It is merely a matter of arrangement and precedence. As for arrangement, it is just as easy and just as neat to slope the numbers towards the left as towards the right. And as for precedence, if a boy has to eat an apple, a pear, and a banana for his lunch, what does it matter whether he starts with the apple or with the banana?

On closer inspection, however, the issue is seen to be more weighty. It is not a question of settling the point of precedence between an apple, a pear, and a banana, but between a melon, an apple, and a grape. Size is an important factor in the problem. The first and largest of the partial products, represented by the three rows of figures, is ten times as large as the second, and a hundred times as large as the third. It may be accepted as a general principle that when we get a result on the instalment plan, as we do in long multiplication, it is sound policy to get the big instalment first—provided there is no reason to the contrary. In addition and subtraction there is reason to the contrary; in multiplication and division there is not.

THE ESSENTIALS OF ARITHMETIC

The traditional mode of procedure seems to derive naturally from the fact that multiplying is a short way of adding. The parallelism of the two processes is manifest in the example 4786 × 3.

Addition	Multiplication
4 7 8 6	4 7 8 6
4 7 8 6	3
4 7 8 6	1 4 3 5 8
1 4 3 5 8	

Here addition and multiplication proceed together step by step, beginning with the units and ending with the thousands. When, however, the multiplier runs into two figures, as in the example 4786 × 13, the parallelism is no longer simple and complete. As far as the multiplicand is concerned, we have to begin with the units as before. Indeed, we are compelled to by the fact that we "carry" from the lower denominations to the higher. But when we come to the multiplier the compulsion no longer applies. We take the addends in bundles, and the more natural thing seems to be to take the bigger bundle first. If we now in imagination convert our first example (3597 × 666) into the two parallel processes, as we have done in actuality in our simpler examples, the reasonableness of taking the largest group of addends first instead of the smallest group is more convincingly brought home to us.

We have still to apply our favourite criterion; we have still to answer the question: Which of the two plans works the better in practice? The reader can

Addition

```
4 7 8 6
4 7 8 6
4 7 8 6
4 7 8 6
4 7 8 6
4 7 8 6
4 7 8 6
4 7 8 6
4 7 8 6
4 7 8 6
```
$\overline{}$
```
4 7 8 6   4 7 8 6 0
4 7 8 6
4 7 8 6   1 4 3 5 8
```
$\overline{}$
```
          6 2 2 1 8
```

Multiplication

```
  4 7 8 6
      1 3
```
$\overline{}$
```
  4 7 8 6
1 4 3 5 8
```
$\overline{}$
```
6 2 2 1 8
```

find an answer for himself by repeating an experiment which I recently carried out in a number of elementary schools. The following set of sums was given to a class to be worked by the traditional method. The children were instructed to work as fast as possible and to give up their papers as soon as they had finished. They were not allowed to correct their errors:

Set I	395867	987563	376859	569837
Multiply	7495	9574	4957	5749

When all had finished, they were given fresh sheets of paper and asked to work Set II by the new method, that is, by beginning with the thousands figure:

Set II	395867	987563	376859	569837
Multiply	5947	4759	7594	9475

It will be observed that the same digits appear in all the examples and that the differences are differences of position only. It will also be noticed that each of the four figures in the multiplier occurs in each of the four possible positions.

I will first give the results for Set I—the set worked in the traditional way. 4436 sums were worked altogether and 5845 errors were made. Eighteen percent of these errors occurred in the first partial product, 24 percent in the second, 29 percent in the third, and 29 percent in the fourth.

The number of Set II sums worked was 4314, and the total number of errors made was 5485. Of these 20 percent occurred in the first partial product, 25 percent in the second, 28 percent in the third, and 27 percent in the fourth.

If in both sets together we take into consideration the first three partial products only, we find the errors distributed among them in the proportion of 26, 34, and 40 percent respectively.

There is one conclusion, of cardinal importance, which we seem to be justified in drawing from these results. It is this: the further a partial product is removed from the multiplicand, the greater the liability to error. And since the first row of figures is the most likely to be accurate, it is better that this row should represent the largest partial product than the smallest. In fine, the newer method of multiplying is better than the older.

Most of the children tested were accustomed to begin at the units end, and it was surprising to me that they so readily adapted themselves to the newer method. In fact the newer method, in spite of its strangeness, proved on the whole the more accurate of the two; for while the average number of errors per sum in Set I was 1.32, in Set II it was only 1.27.

It will be observed that the distance of a partial product from the multiplicand may be estimated in two distinct ways—laterally and vertically. In multiplying by the units figure there is no lateral dislocation; each figure of the product is placed directly underneath the corresponding figure in the multiplicand. In multiplying by the tens there is a dislocation of one place, in multiplying by the hundreds a dislocation of two places, and so forth. I expected to find this lateral displacement affecting the partial products much more than it actually does. It does apparently disturb them slightly, as is seen by the greater degree of accuracy in the first partial products of Set I than in the first partial products of Set II. The disturbance due to lateral distance is, however, negligible in comparison with the disturbance due to vertical distance.

The upshot is a triumphant victory for the newer method.

CHAPTER XIII

MULTIPLICATION OF MONEY

A great reckoning in a little room.

SHAKESPEARE: *As You Like It.*

THERE is probably no single process in the whole realm of arithmetic which is put to such constant use in the workaday world as the multiplication of money. In buying and in selling, in travelling and in sight-seeing, in going to the play or to the restaurant (I assume we don't do these things alone), we are always calculating costs by multiplication. And although the process does not loom so large in textbooks as in life, yet much of the space in our textbooks is devoted directly or indirectly to the multiplication of money. Practice, both simple and compound, is but another way of multiplying. Indeed, it used to go under the name of Small Multiplication. Money is sometimes decimalised in order that it may be more easily multiplied, or may more readily enter into the substitute process of practice.

It is therefore of cardinal importance that our standard rule for the multiplication of money should be a good one. In searching for a good rule we are confronted with many difficulties. The first is that the task itself is difficult. There is a large variety of rules in use, and none of them is easy. A well-known administrator in London was wont, on his visits to elementary schools, to set the pupils in the top class to work some such example as this: £5 16*s.* 7½*d.* × 29. He generally got deplorable results. It wasn't that the pupils couldn't do it, but that they couldn't do it correctly. Little mistakes crept in at all stages. An official investigation was made into the standard of attainments in the elementary schools of London in 1924. Here are two examples from the arithmetic test:

(a) How much would 92 times £53 8*s.* 3*d.* be?

(b) £45 11*s.* 5½*d.* is divided equally among 25 men. How much does each get?

It is generally believed that long division of money is more difficult than the multiplication of money, yet the children did the second sum much better than the first. While only 60 percent of the boys and 58 percent of the girls got the multiplication right, 73 percent of the boys and 70 percent of the girls got the division right. There is no doubt whatever about it: children in our best schools are peculiarly liable to get muddled over compound multiplication.

The methods in use, varied as they are, may yet be regarded as falling broadly into two classes, which may, for convenience' sake, be called piecemeal and wholesale. By a piecemeal method I mean one which involves either factorial multiplication or the addition of partial products. If I multiply a sum of money by 18 by multiplying double the sum by 9, I am using a piecemeal method. If I take ten times the sum and add to it eight times the sum, I am using another kind of piecemeal method. To multiply by, say, 284, a combination of these two types is generally adopted, the steps depending upon the fact that $284 = (10 \times 10 \times 2) + (10 \times 8) + 4$. It is the commonest of the piecemeal methods, and the procedure and arrangement need no illustration.

In the wholesale method, on the other hand, the multiplier is taken as a whole and the product is given in one line, as though the multiplier were a single digit. In working out the details, differences and divagations abound. Some do it in the margin (and generally get tangled up); others run a long serpentine column down the page. Some begin with the pence, some with the pounds; some introduce practice methods, others confine themselves to pure multiplication. The one point in which they all agree is that they finish dealing with one denomination before they proceed to the next.

In my quest for a sound standard method my first task was to discover which of the two broad systems proved the more efficient in practice. It did not take me long to gain a strong impression that the wholesale method was both the more accurate and the more speedy. Let me cite an example of the kind of evidence on which my opinion was based. I set the following sum to a class of 25 girls in Standard VI, some of whom had been accustomed to work by a piecemeal method and some by a wholesale method: £53 14s. 8½d. × 47. Out of the 12 who used the wholesale method, 11 got the sum right; out of the 13 who used the piecemeal method, 3 got it right. My next task was to find the best of the wholesale methods. After observing several methods in operation and giving a few tests, I came to the conclusion that the best is the one illustrated here under the name of the standard model.

£	s.	d.	
58	14	8¾	
		365	
21438	16	1¾	f.
268	266	273	4)1095
18250	3650	2920	273¾d.
2920	1460	12)3193	
	20)5376	266s. 1d.	
	£268 16s.		

<small>STANDARD MODEL IN FULL</small>

In teaching this rule, the first step is to practise the children in multiplication where the multiplier, instead of being in its usual position, is above the multiplicand. This need not take up much time. As a matter of fact it rarely matters where the multiplier is placed: the pupil rarely looks at it. He generally carries it in his head. The next step is to practise multiplying money of one denomination; e.g. 5d. × 48 (which, by the law of commutation is equivalent to 5 × 48d.); 7s. × 250; 16s. × 192, etc. Then the full process should be explained and a worked model left in view of the class for purposes of reference.

The farthings will present a slight difficulty. It is better to avoid giving them a separate column. This is

£	*s.*	*d.*
58	14	8¾
		365
21438	16	1¾
268	266	91¼
18250	3650	273
2920	1460	2920
	5376	3193
	268	266

STANDARD MODEL IN
ITS NEATEST FORM

easily done by taking the farthings frankly as they are written—as fractions of a penny and, as such, belonging to the pence column. The ¾*d.* is the least easy to deal with. It may be given a column of its own, as in the full form of the standard model, or be dealt with wholly under pence, as in the second form of the model. If, for instance, the pence to be multiplied in the model were 8¼*d.* instead of 8¾*d.*, the first row of figures underneath would be 91¼, or, preferably, 91 only, with the ¼ put straight down in the answer. The 91¼ is, of course, 365 ÷ 4. If the pence had been 8½*d.*, the first row would have been 182½, or 182. As it is 8¾*d.*, use should be made of the fact that ¾*d.* is 1*d.*— ¼*d.*, and 91¼ should be subtracted from 365.

The abbreviated form of the model is the one to work up to. All superfluities are omitted. The dividing numbers, 4, 12, and 20, are kept in the head, and the remainders are put down direct in the final product. The advantage of this piece of economy is that it prevents the three columns from barging into one another—a thing they are liable in any case to do unless £ *s.* and *d.* are spread as far apart as the paper will admit. It should therefore be a standing instruction to the pupils to begin by spreading the three denominations widely across the page.

If the work is kept neat and orderly (scribbling is fatal to accuracy, especially if it is cramped into a narrow space such as the margin), it will, I think, be found that this is the most effective mode of multiplying. How do I know? Because I have tested it over and over again. My first victim was myself. I had no special predilection for any method and had used several indiscriminately. I worked a large number of examples by both the wholesale and the piecemeal methods, timing myself with a stop-watch, and found that in both speed and accuracy the advantage lay distinctly on the side of the wholesale method.

It by no means followed that because I myself worked better by the wholesale method, others would work better too. And it is specially unsafe to assume that what is true of adults is true of children. I therefore proceeded to test the children in elementary schools. The tests were as a rule, though not always, given to the highest class, and generally consisted of the same sums to be worked by the two rival methods. Here is one type of experiment. A class of children sitting in dual desks is divided into A's and B's, those sitting on the right side of the desks being labelled A's, those on the left, B's. A sum placed on the blackboard has to be worked in the wholesale way by the A's, and in the piecemeal way by the B's. Then another sum of about equal difficulty is placed on the blackboard, and the methods of working changed round. Those who worked the first sum by the wholesale method have to work the second sum by the piecemeal method, and *vice versa*. The time taken by each pupil in working each sum is estimated in the manner described on pp. 132-133. The results varied somewhat from class to class; but the wholesale method always took less time and nearly always proved the more accurate.

I can best convey an idea of the kind of data I collected by giving a specific example. I visited a certain boys' school in a poor district and proposed to test the top class. The headmaster held very strong and very definite views on the subject under inquiry. He disliked the wholesale method and had banished it from his school. It was necessary, therefore, for the purpose of my test to explain to the class what the wholesale method was. I illustrated it by working an example on the blackboard: £53 14s. 8½d. × 47. The boys were then set to work the following example by themselves: £8 13s. 5¼d. × 365. When they had finished, their papers were collected, fresh papers were given out, and the class was asked to work the same example by their customary method—the piecemeal method. It will be seen that the advantage in point of familiarity was all in favor of the piecemeal method, as also was the advantage in order of testing. It came second; it came, after one attempt had already been made at the sum. And yet

9 boys got it right by the wholesale way and only 8 by the piecemeal. The average time taken by the 9 boys was 3.25 minutes, and by the 8 boys 4.25 minutes. Taking the class as a whole, the average times were 3.7 and 4.3 minutes respectively. They therefore worked by the unfamiliar method better than by the familiar method.

Rarely was the evidence quite so strong as that; but it always pointed to the wholesale method as the superior method. Whenever the wholesale method seemed to fail, the failure was due to a lack of order and neatness in the working rather than to any inherent defect in the method itself.

I occasionally gave the same examples to be worked in two other ways—by practice and by decimalising the money. The outcome was always disappointing: an inordinate time was taken in producing the most meagre and unreliable results.

It is scarcely necessary to say that practice should be regarded as an alternative mode of multiplying—sometimes a better mode, sometimes a worse. I well remember my young days when I gained my first acquaintance with practice. I learnt it as a distinct and independent rule. It was the rule for working sums with two distinguishing marks, "articles" and "@." All things created by God or man were called "articles," and the plain preposition "at" was given an ornamental tail. If we were asked to find the cost of 450 articles @ £1 18s. each, we knew we had to work by practice; if to find the cost of 450 clocks at £1 18s. each, we knew we had to multiply.

Practice methods are available at all stages of a multiplying process; it can be inserted at any point and to any degree. If, for instance, the multiplicand is £15 16*s*. 6*d*., it would be sheer waste of time to use a pence column, for the money may much more conveniently be regarded as £15 16½*s*. A precaution has to be taken. In dividing the multiplier by 2, the remainder, if there is a remainder, is 6*d*., not 1*d*. Indeed, the more we introduce short and "intelligent" methods into routine sums, the wider do we open the door for errors to enter. This is no argument for extruding intelligent methods: it is merely a counsel to caution. If you use intelligent methods, check your result by other intelligent methods—or by the routine method.

CHAPTER XIV

MULTIPLICATION OF DECIMALS

To fetter reason with perplexing rules.

POMFRET: *Reason.*

In my young days the multiplication of decimals was quite a simple matter. We ignored the decimal points, multiplied, and then counted the places. Sometimes the reason was given; generally it was not. In any case, we did not clamour for a reason, but worked the sum according to rule and were happy. Nowadays the matter is not so simple. Children are required to multiply by a variety of rules, none of which is as easy to apply as the old-fashioned rule, but each of which is alleged to have peculiar merits of its own. The new rules are perplexing, but beneficent. So it is claimed.

The old method is one and indivisible, the new method splits into many varieties. The old is based on the fact that a decimal fraction is a fraction, the new on the fact that it is a decimal. They thus result from two different modes of approach, one of which may be called the fractional approach and the other the notational approach.

Let us first consider the fractional approach, which is also the traditional approach. The pupils are supposed to be familiar with vulgar fractions, and to be able to multiply them; they are supposed to know the rule that we multiply the numerators together to get the new numerator, and multiply the denominators together to get the new denominator. That is the groundwork on which we build. We proceed by some such steps as these. Suppose we have to multiply .5 by .07. Written as vulgar fractions they are $\frac{5}{10}$ and $\frac{7}{100}$, and $\frac{5}{10} \times \frac{7}{100} =$

$\frac{5 \times 7}{10 \times 100} = \frac{35}{1000}$, which equals .035.

Let us now try mixed numbers, such as 3.284 x 13.6. Worked as vulgar fractions this expression becomes

$$\frac{3284}{1000} \times \frac{136}{10} = \frac{3284 \times 136}{1000 \times 10} = \frac{446624}{10000}$$

This product equals 44.6624. Place the results together thus:

$$.5 \times .07 = .035$$
$$3.285 \times 13.6 = 44.6624$$

Now ask the class to formulate a general rule. If they fail to do so, give a few more examples. They will at once see how to find the significant figures in the product; the sole difficulty is in fixing the decimal point. In the instances given, the number of decimal places in the product is equal to the sum of the number of places in the factors. Does the rule apply universally? It is easy to show that it does.

The following facts should be clearly understood:

$$\frac{1}{10} = \frac{1}{10^1} = .1$$

$$\frac{1}{100} = \frac{1}{10^2} = .01$$

$$\frac{1}{1000} = \frac{1}{10^3} = .001$$

The index of 10
= the number of naughts
= the number of decimal
places.

$$10^2 \times 10^3 = (10 \times 10) \times (10 \times 10 \times 10) = 10^{2+3} = 10^5$$

From these and similar examples the brighter children at least may be brought to understand that if A and B represent two decimals of m and n decimal places respectively, their product is obtained thus:

$$\frac{A}{10^m} \times \frac{A}{10^n} = \frac{AB}{10^{m+n}}$$

That is, the number of decimal places in the product of A and B is $m + n$. I do not imagine the beginner will grasp the principle so as to be able to express it, or even to understand it, in this symbolic form; but I do think he can be brought to realise that the procedure prescribed is a reasonable procedure, and that when he is multiplying the numbers he is multiplying numerators, and when he is fixing the decimal point he is multiplying denominators.

It will thus be seen that the traditional method is logically coherent, is by no means difficult to understand, and is delightfully easy to apply.

Many English mathematicians, however, set their face against this mode of attack. They hold that instead of deriving our method indirectly through vulgar fractions we should derive it directly from our decimal notation of whole numbers. Just as we teach our pupils to add and subtract decimals without any reference to corresponding operations with vulgar fractions, so should we teach the multiplication and division of decimals. Let us see how this may be done.

H. OF TH.	T. OF TH.	TH.	H.	T.	U.	
			4	9	6	$\ldots a$
		4	9	6		$\ldots b = a \times 10$
	4	9	6			$\ldots c = b \times 10 = a \times 10^2$
4	9	6				$\ldots d = c \times 10 = b \times 10^2$
						$= a \times 10^3$

In the above illustration imagine the places fixed and the figures movable. To multiply a number by 10 we move all the figures bodily one place to the left; to multiply by 100, or 10^2, we move them two places to the left, and so on.

Consider this example, in which the above number, 496, is multiplied by 257, and the partial products are shown to be derived from a, b, c.

$$
\begin{array}{rrrrrr}
 & & & 4 & 9 & 6 \\
 & & & 2 & 5 & 7 \\
\hline
 & 9 & 9 & 2 & & & \ldots\ c \times 2 \\
 & 2 & 4 & 8 & 0 & & \ldots\ b \times 5 \\
 & & 3 & 4 & 7 & 2 & \ldots\ a \times 7 \\
\hline
1 & 2 & 7 & 4 & 7 & 2 \\
\end{array}
$$

We shift all the figures bodily and then multiply by the appropriate digit, except in the last partial product, where the figures remain *in situ*.

Now extend the system below the decimal point.

Let t represent tenths, h hundredths, etc.

H. T. U.	t. h. th.	
4 9 6		$\ldots a$
4 9 . 6		$\ldots p = a \times .1$
4 . 9 6		$\ldots r = p \times .1 = a \times .01$
. 4 9 6		$\ldots s = r \times .1 = p \times .01 = a \times .001$

The following example may now be studied—an example with an exceptionally large number of figures in order to illustrate a principle:

```
        8 3 . 7 5 6
        3 7 [2] . 4 9
```
2 5 1 2 6 . 8	= 300	times the multiplicand
5 8 6 2 . 9 2	= 70	" " "
1 6 7 . 5 1 2	= 2	" " "
3 3 . 5 0 2 4 =	.4	" " "
7 . 5 3 8 0 4 =	.09	" " "
3 1 1 9 8 . 2 7 2 4 4 = 372.49		" " "

The point of reference is the units figure of the multiplier; i.e. the figure 2 in the above example. By studying the examples given we arrive at this general rule. The figures forming the multiplicand remain in position when we multiply by the units figure. When we multiply by any other figure the figures of the multiplicand move to the left or the right according as the multiplying figure is to the left or the right of the units figure. The number of places the figures are shifted is determined by the number of removes of the multiplying figure from the units position.

All this sounds very complicated, but it can be rendered quite simple by blackboard illustration and a few brief questions and exercises.

A great advantage is gained if the units figure of the multiplier is placed, not under the units figure of the multiplicand, but under the last decimal figure. Thus:

```
        8 3 . 7 5 6
          3 7 [2].4 9
  2 5 1 2 6 . 8
    5 8 6 2 . 9 2
      1 6 7 . 5 1 2
        3 3 . 5 0 2 4
          7 . 5 3 8 0 4
  3 1 1 9 8 . 2 7 2 4 4
```

The two main merits of this arrangement are: first, that each partial product begins under its own multiplying figure (which brings it into line with the rule for simple multiplication), and secondly that the number of decimal places can be added at a glance (which brings it into line with the traditional rule for multiplying decimals).

But matters have not been allowed to rest even there. Another method has come into vogue of late years—the method of "standard form." By this method we must juggle with the figures before we begin to multiply. But perhaps I had better, for the behoof of the uninitiated, explain what standard form means. A number is expressed in standard form when it is written as a mixed decimal with a single digit before the decimal point multiplied by a positive or negative power of ten. For example, 16,540,000 expressed in standard form is 1.654×10^7, and 0.0005187 is 5.187

126

$\times 10^{-4}$. It is a convenient way of exhibiting the order of magnitude. It is not easy at a glance to say which is the larger, 0.000000013 or 0.0000000013. But when these are written in the form 1.3×10^{-8} and 1.3×10^{-9}, it is at once seen that the first is ten times as large as the second. But, it will be objected, we never in ordinary life deal with such small quantities. In science we do. In dealing with the dimensions of atoms and electrons, or in finding the time it takes light to traverse small distances, we have to deal with quantities extremely small. The standard form is equally convenient for expressing such large numbers as are required in estimating the distances of the stars or the frequency of a wireless wave or the number of electrons in a given space.

There is another, and more important, use of standard form. All the logarithms given in logarithmic tables are logarithms of numbers in standard form. I open a book of such tables at random and find that the logarithm of 41632 is 6194273. This means that the logarithm of 4.1632 if .6194273. From this I may deduce the logarithm of any number which has 41632 as its significant figures. For example: 416.32 is 4.1632 $\times 10^{2}$, and its logarithm is therefore 2.6194273.

The reader is by this time wondering what all this has to do with the multiplication of decimals. Well, many mathematicians advocate the conversion of the multiplier into standard form before beginning to multiply. They contend that it is an easy way of securing an approximate answer, an easy way of fixing the decimal point, and that it familiarises the pupils with a notion which will prove useful to them when they come to deal with logarithms. Suppose, for instance, we have to multiply 175.36 by .264. We must first turn .264 into standard form by multiplying by 10, and compensate for this by dividing 175.36 by 10. Thus:

$$175.36 \times .264$$
$$= 17.536 \times 2.64$$

It will be observed that if we move the decimal point in one direction in one of the factors we have to move it the same number of places in the opposite direction in the other.

So far all who adopt the standard-form method agree. The next step is to find a rough approximation; and here disagreement begins to creep in. While some would take 2 as the multiplier and give 35 as the approximate answer, the majority would, more wisely, take 3 as the multiplier and regard 52.5 as an approximate result. Others again would take limits and argue thus:

The answer cannot be less than $17 \times 2 = 34$

The answer cannot be greater than $18 \times 3 = 54$

The answer therefore lies somewhere between 34 and 54.

At this stage there is a sharp divergence. One faction henceforth ignores the decimal point until the end is reached; the other regards the ignoring of the decimal point at any stage as a cardinal offence. The first faction, treating the two factors as whole numbers, obtains 4629504 as a product. As the rough approximation shows that there are two figures in the integral part of the answer, the final result is written 46.29504. In fact the procedure differs from the traditional rule by counting from the left instead of counting from the right, the number of figures to be counted being given in one case by the rough estimate and in the other by the sum-total of decimal places.

The other faction, after finding the rough approximation, proceeds to work the example according to one of the two methods illustrated on pp. 125-126. The precise answer, with the decimal point correctly placed, is obtained by multiplication, the rough approximation being used as a check only.

I thought I knew the extreme limit to which a fervour for standard form could carry a teacher. But I was wrong. Since writing the above I have found a mathematical master who insists upon his pupils putting *all* decimals into standard form before multiplying or dividing. The proper way to multiply 148.36 by .0057 is this: $148.36 = 1.4836 \times 10^2$, and $.0057 = 5.7 \times 10^{-3}$. The required product therefore is $1.4836 \times 5.7 \times 10^2 \times 10^{-3}$. The first two factors, multiplied by the usual notational method, come to 8.45652. But $10^2 \times 10^{-3} = 10^{2-3} = 10^{-1}$. Therefore the product already found has to be divided by 10, and the final answer is .845652. I have not tested this man's pupils, but I should very much like to. I should also like to know what they think about it. As a new method it ranks with that of the boy who had discovered a new way to count sheep: count their legs and divide by four.

CHAPTER XV

A TILT AT STANDARD FORM

> It is true that the patient died under the treatment, but we have the consolation of knowing that he died cured.
>
> KEILEMOWSKI: *Lectures on Experimental Medicine.*

WE have seen that there are two distinct ways of multiplying decimals—the fractional and the notational—and that there are two or three varieties of the notational method. In actual practice, however, the only variety that counts is that of "standard form." Except in rare instances, no other variety is found in our schools. The real fight, therefore, is between the traditional method on the one hand and the standard-form method on the other. And before taking sides and arguing the matter from first principles, let us take a steady look at the combatants and see how they acquit themselves in the arena. Let us, in fine, apply the rough-and-ready criterion we have always adopted, and ask the simple question: Which is the better method for getting sums right?

As for those who adopt a lofty tone and refuse to accept this criterion, hinting vaguely at broader aims and higher issues, I will for the moment content myself with commending to their notice a certain incident which occurred when a visit was paid some years ago to a small nursery school. It was observed that the children were put to sit at their little tables and given a mid-morning cup of milk, the milk standing scarcely higher than an inch in each cup. When the visiting doctor suggested a more generous supply he was met with this retort: "You have not grasped the idea. The purpose of giving these children milk is not that they should be nourished, but that they should learn table manners." Those who hold that the purpose of teaching children to multiply decimals is not to enable them to multiply decimals but to teach them something else will take no interest in the evidence that follows, and would be well advised to read no further. Those, on the other hand, who agree with me that the only measure of the worth of a method is the success with which it achieves its avowed purpose will regard the evidence given below as the only sort that is of any consequence.

Broadly speaking, the traditional method is taught in elementary schools and the standard-form method in secondary schools. The division, however, is not sharp and clean-cut; for in some parts of the country the elementary schools, influenced by the secondary schools to which they contribute, try to teach standard form, and some secondary schools (and, judging from examination results, some of the best secondary schools) adopt the traditional method.

I discovered that the best place to test the comparative efficacy of the two methods was the third or fourth form in a secondary school. I accordingly tested a number of such classes in various secondary schools, some of them boys' schools and some girls'. The ages of the pupils ranged from 12 to 15. Since entering the secondary school they had all practised the standard-form method only, but the majority of them knew the traditional method as well, for they had learnt it at their previous schools. Paper A was distributed, and the pupils were asked to work the sums as rapidly as possible by the standard-form method and to record the time taken:

A	B
(1) 41.76 × 20.3	(1) 23.81 × 30.2
(2) 314.2 × .18	(2) 142.3 × .17
(3) 48.16 × .072	(3) 56.18 × .081
(4) 730.5 × 321	(4) 620.4 × 132

The mode of estimating and recording the time was quite simple. I stood near a blackboard with a stop-watch in one hand and a piece of chalk in the other. I indicated the time as it passed, writing on the blackboard at intervals of 15 seconds the number of minutes and quarter-minutes. Thus the following numbers appeared successively on the blackboard: 2, 2¼, 2½, 2¾, 3, 3¼, 3½, 3¾, 4, etc. Only one of these was visible at any time on the board, so that all the pupil had to do when he had finished his sums was to look at the blackboard and record the time he saw written there.

Before giving the second test I devoted a few minutes to refreshing the memories of the majority who had practised the traditional method in past years, and in explaining the rule to the few who had never practised it. Then paper B was distributed and the pupils asked to work it by counting the decimal places. The time was recorded as before.

The two tests are, as will be observed, of about equal difficulty. To neutralise any advantage which one of them might possibly have over the other I changed them about, giving A to be worked by the standard-form method in some schools, and B in other schools. It made no difference to the results: in both cases they told the same tale. They are set forth in Table 1:

School	Pupils	Standard Form		Traditional Method	
		Sums	Minutes	Sums	Minutes
M	33	2.4	5.2	3.1	3.1
N	27	2.8	5.5	3.4	3.8
O	28	2.6	4.2	3.2	3.7
P	30	2.2	5	3.4	2.8
Q	21	3	4.5	3	3.4
R	22	3	3.7	3.3	1.8
S	25	2.1	5	2.9	3.1

TABLE 1

Table 1 should be read thus. In school M 33 pupils were tested. By the standard-form method an average of 2.4 sums were got right in an average time of 5.2 minutes. By the traditional method the average number of correct sums was 3.1 and the average time taken was 3.1 minutes.

I also used another set of tests, C and D, for the same purpose; but as the advocates of standard form regarded them as weighted against their method, they were used less frequently than the A and B tests.

C		D	
(1) 107.03	× .12	(1) 205.01	× .011
(2) .005	× .0007	(2) .009	× .0005
(3) 5.006	× 10.3	(3) 4.007	× 10.4
(4) .34	× .34	(4) .26	× .26
(5) 26.35	× 21	(5) 17.24	× 21
(6) .0175	× 110	(6) .0164	× 120

TABLE 2

The results for C and D are given in Table 3.

School	Pupils	Standard Form		Traditional Method	
		Sums	Minutes	Sums	Minutes
S	32	3.4	6.7	4.1	4.4
T	23	5	9.1	5.3	5

TABLE 3

It is abundantly clear that the standard-form method is less accurate than its rival and takes more time. The number of sums worked correctly in a given time is about half as great. The apologists for the method contend that it takes more time because the pupils have to make a rough estimate before they begin the multiplication proper. But I found on examining the papers that only a small number of the examinees, sometimes only two or three in the whole class, attempted to make a preliminary estimate. They changed the multiplier to standard form and began multiplying straight off. And taking the papers one by one, and comparing the times taken by the competing methods, I found the advantage of speed to be invariably on the side of tradition. The extra time involved in the longer method was apparently spent in thinking out the reduction to standard form, and in keeping the decimal point in its right position in the partial products. Its wastefulness of time, in fact, is due to its own inherent qualities as a method of multiplying, and cannot be ascribed to any adventitious benefit which is alleged to accrue.

The source of its inaccuracy is not far to seek: it is the first step—the conversion to standard form. The errors due to mistakes in multiplying or carrying are pretty evenly balanced between the two methods (though even here the traditional method has a slight advantage); it is in the characteristic feature of the method—the initial adjustment of the figures—that errors greatly abound. And the resulting blunders are not of the venial kind, where the order of magnitude is right and a digit only is wrong, but of the flagrant and unforgivable kind where the decimal point is grossly misplaced. Here is a typical example:

$$41.76 \times 20.3$$
$$\text{S. F.} \quad 4.176 \times 2.03 \qquad \text{R.A.} = 8$$
$$\begin{array}{r} 4.176 \\ 2.03 \\ \hline 8.352 \\ 12528 \\ \hline 8.47728 \end{array}$$

The error here seems due to confusion with the compensation to be made when the divisor of a division sum is converted into standard form. Whatever the cause, it is the most frequent error, and results in the answer being a hundred times as large or a hundred times as small as it ought to be. That is the smallest magnitude of the error, as it results from shifting the decimal point one place only. When the point is shifted two places the answer becomes ten thousand times as large or as small as it should be. And the rough approximation doesn't save the sum. It merely corroborates the sum's own lie.

It is well, therefore, that we should examine pretty closely the claims made on behalf of the method of standard form. It is first of all claimed that it prepares the way to the understanding of logarithms. It does this apparently by familiarising the pupils with the notion of standard form; it makes them associate the terms with a mixed decimal whose value lies between 1 and 10. Perhaps it does more than this. Perhaps it makes them dimly realise that the same significant figures may represent a variety of magnitudes all resembling one another in having a common factor, and all differing from one another by having a variant factor which is ten or some power of ten, and thus helps them ultimately to grasp the fact that all numbers with the same significant figures have logarithms with the same mantissæ. Perhaps it does this. But is it necessary to confuse and bedevil the whole of the plain process of multiplying decimals to secure so dubious, so hypothetical, a benefit? In actual practice the method seems to resolve itself into a mechanical shifting of points. Not that I object to it simply as a piece of mechanism; but as a piece of mechanism which is peculiarly liable to go wrong, and which pretends to be something more than it really is.

The next claim is that it clears the ground for contracted methods. True; but so does every other mode of notational approach, and does so equally well. The method of standard form has no advantage in this respect over the methods illustrated on p. 126. The only serious obstacle to a ready grasp of contracted multiplication is the habit of starting to multiply with the units figure of the multiplier. But that habit is gradually disappearing from our schools. Children now, in ever-increasing numbers, begin to multiply from the weightier end, and thus secure at once the essential principle of contracted methods; and secure it without being indebted to any adventitious source.

Even if we admit that the method of standard form does in a measure facilitate the future teaching of logarithms and contracted methods, are we on that ground alone wise to introduce the method, with all its imperfections thick upon it, to the multitudes of young innocents in our schools? Of the whole school population, primary and secondary, less than 5 percent will ever need logarithms or contracted methods, even for examination purposes. For purposes of social or business life the numbers are still smaller. Are we then justified in imposing upon all and sundry a method which is a real and present nuisance to the many, on the score of its being a future and problematic benefit to the few?

Let us, however, proceed to examine the further

claims that have been made on behalf of the method. It is constantly being put forward as the best way of arriving at an approximate answer. But what is the use of an approximate answer except as a check upon the final answer? Unless it serves as a safeguard against an absurd result it is nothing but a meaningless piece of ritual, of no more value than a rough estimate of a friend's telephone number made before looking it up in the telephone book. The rough guess in the standard-form method, however, fails as a check on the final result, not because there is no connection between the check and the result, as in the telephone number, but because there is too much connection. As sponsors for the truth they are not independent witnesses: they are in open collusion. Whatever one says the other will corroborate it. Look at the example on p. 135 and see the conspiracy at work. The rough approximation is made too late to be a genuine confirmation or refutal of the final verdict. To be a genuine check it must be based upon the original unsophisticated figures. Let us consider the example in question, 41.76×20.3. It is clear to the meanest intelligence that this is roughly 42×20, or 840, which is a closer approximation than is usually reached by standard-form methods. And whatever may be the method by which the example is worked out in full, this rough estimate will stand as a trusty criterion of its essential rightness. The common-sense method of getting a rough result can easily be applied to nearly all examples. Let us try it on the other examples in Test A. The second sum is $314.2 \times .18$.

As .18 is obviously between ⅙ and ⅕, the answer lies between 50 and 60. The third sum is 48.16 × .072. Since $7/100$ is about $1/15$, the answer is about 3. The fourth sum is 730.5 × 321. The product is about 800 × 300, that is about 240,000. A disquieting percentage of pupils who worked this last example by standard form and made a rough guess gave 21 as the guess and justified it by a corroborative answer.

Some of the advocates of standard form frankly admit that the rough answer is not used as a check but as a means of determining the integral part of the product—as a means, in fact, of fixing the decimal point by counting from the left just as the traditional method fixes the point by counting from the right. The counting, however, is sometimes precarious. Can we say straight off, by a simple standard-form guess, how many figures there are in the integral part of this product: 55.4 × .18?

The final refuge of the apologists for standard form is the remark: "At any rate it makes the pupils think." It does. That is its most serious indictment. It wastes mental effort. It is mentally laborious without being mentally efficient. If it were an intelligence test, the plea would be relevant and valid, for its essential aim would be to provoke thought and to measure it; but as it is a routine sum with no element of novelty, the real demand is for efficiency—for a frictionless and trustworthy turning out of the answer. As Sir Oliver Lodge puts it in discussing another rule: "It is at best a mechanical process, and it should be done mechanically; that is by a straightforward method which involves no delicate thought, and affords no loopholes for mistakes to creep in." [18]

The inaccuracy and tardiness of the standard-form method are matters which admit of no doubt and no extenuation. The results which I have recorded cannot be explained away by assuming that the rival method had been well drilled into the testees when they were in the elementary schools. It is true that a large proportion of the pupils were scholarship children; but they had taken the scholarship examination before they were eleven years of age, and it was an examination which laid but slight stress on decimals. In their brief and crowded school career there had been no time for drill in decimals. And as a matter of fact it made no difference whether the examinees had just entered the secondary school or had been there for three or four years: the superiority of the traditional method was equally manifest in both cases. Indeed the minority who had picked up the traditional method a few minutes before the test, if they had rightly grasped the rule, worked by this method with greater facility than with the other. The truth is that as far as familiarity was concerned it was the standard form that had the advantage. It had the great advantage of recent practice, for the other method had often been tabooed for years.

The testees were mainly scholarship children, children with keen intellects, children who were mentally two or three years in advance of their age. And if they found difficulties with the standard-form method, what would have happened to their less

[18] *Easy Mathematics,* p. 232.

intelligent comrades whom they had left behind in the elementary school? And yet in two counties at least (and probably many more) attempts had been made to enforce the standard-form method on all the elementary schools within their borders. As a rule the pressure originally emanates from one or two brilliant mathematicians in secondary schools who have themselves been nourished on standard form, who exaggerate its importance as part of a mathematical training, and are eager to secure a few scholars whom they can pass through the same mathematical mill as themselves. And to do this they are willing to sacrifice the serenity of a large multitude of ordinary human boys and girls to whom the higher mathematics will ever be about as interesting and intelligible as Hegelian metaphysics.

To pooh-pooh inaccuracy and failure—to assert that these things are of little consequence for after-school life—is to miss the important moral training that mathematics brings. Who is riding the high horse now? the reader will no doubt ask. Well, I am. I don't go so far as to say that we should teach mathematics in order to inculcate morals (my real belief is that we should teach mathematics in order to inculcate mathematics, and that if we do it well other boons are added gratis), but I do definitely affirm that we should sedulously foster a respect for truth, that we should keep the ideal of accuracy burning clear and bright, that we should cultivate that confident attitude of mind which makes for success, and that we should above all things avoid developing a lack of confidence and a permanent expectancy of failure. I think it morally corrupting to tempt children to get sums wrong wholesale; and no ulterior benefit, even if it were far less hypothetical than the benefit usually offered, can justify the corruption. The roads that lead to knowledge are rough enough in all conscience, and no wise teacher will make them rougher than they need be.

1. Let me now come down from my high horse and return to decimals. It will be observed that my tests have consisted of two factors only. What happens when these are more than two—when the pupils are set to find the value of, say, $.02 \times .6 \times .34$, or of $.008^3$, or of $11.3^2 \times .36^2$? The more rigid of the sect of standard-formers would forbid all counting of decimal places (except on their own method) and would, I presume, solve the first example like this: $.02 \times .6 = .002 \times 6 = .012$; $.012 \times .34 = .0012 \times 3.4$, which would, if the standard-form routine were strictly adhered to, be worked in two lines; as indeed it would if it appeared in the form $.0034 \times 1.2$. How much easier to say: $2 \times 6 = 12$; $12 \times 34 = 408$; and as there are five decimal places altogether the answer is $.00408$!

To discard altogether the old-fashioned rule, even as an auxiliary or as a check, is to reject the services of a valuable friend. Even the rule for shifting the decimal place so as to compensate for the change into standard form—the process which the pupil finds so puzzling—would lose its peculiar liability to error if the pupil always asked himself the question: Have I left the sum-total of decimal places unchanged? If he realises that he may juggle with the points in any way he likes so long as he does not alter the total number of places, he can deal with the above example in summary fashion thus: .02 × .6 × .34 = 2 × 6 × .00034 = .00408.

Having shown up the weakness of the method of standard form, let me now exhibit the strength of the traditional method. In the first place it prepares the way for a study of logarithms. For if it is properly taught it is seen to derive from the primal fact that $a^m \times a^n = a^{m+n}$, which is the fundamental principle upon which all operations with logarithms ultimately depend. In adding decimal places to fix the decimal point we are adding indices. In adding logarithms we are adding indices. In both cases we are multiplying numbers by adding other numbers related to them by the law of indices. This is not merely a useful convention like standard form, but is of the very essence of logarithmic theory.

When the method does not actively help, it at least does not hinder. Although it does not itself permit of contracted process, it leads naturally to a method (see p. 126) which readily permits of contraction. Although it does not offer an approximate answer which serves as a check upon the final answer, it only fails to do what every other method fails to do. And at least it does not betray the pupil into accepting a spurious check.

The fact that the rule is easy to remember and can be applied with the minimum expenditure of thought and volition—a characteristic which some have regarded as its gravest fault—should, for reasons which I have abundantly expounded, be regarded as its crowning grace.

It is alleged that the traditional rule is not only applied mechanically (which is just how it should be applied) but learnt mechanically too. If the charge is true there is admittedly a defect, though not a serious defect. But the defect is not in the rule: it is in the teaching. The rule is neither difficult to teach nor difficult to understand. Even when the pupil fails to follow the reasoning he at least acquires a confidence in the reasonableness of the rule. And this is something to go on with at any rate. He is in no worse case than the ordinary man who works square root by rule of thumb.

I believe that the idea of figures moving to the left or right in accordance with the denomination of the multiplying digit is one of peculiar difficulty to children. If the movement can be made by a straightforward mechanical rule, as in simple multiplication, the liability to error is small, but when it is a sort of see-saw movement, as in the multiplication of decimals by the standard-form method, I believe the child has the same kind of mental bewilderment as the adult who tries to think out whether he has to move the hands of his watch an hour forwards or an hour backwards when the change is made from Greenwich time to summer time. By some means or other, however, by mechanical means or by rational means, multiplication must sooner or later be attacked from the notational side. I myself think that the fractional method, being easier and more universally useful, should be taught first, and the notational method, preferably the second form (see

p. 126), taught afterwards. This notational method is sometimes particularly apt. If, for instance, we wish to find 6 percent of £276, the simplest way is this:

£276

£ 16.56

We multiply the 276 by .06, beginning two places to the right, and thus of course moving all the figures two places to the right. If we wish to repeat the process, as in finding compound interest, we again multiply by six by moving two more places to the right.

My final counsel is that the traditional rule for the multiplication of decimals should be adopted as the standard rule. I, however, make these provisos: that multiplication begin from the left of the multiplier, that the rule be logically connected with the multiplication of vulgar fractions, and with the general law of indices, $a^m \times a^n = a^{m+n}$, and that some form of notational method be taught as well.

There is one point of interest (and of significance) which I have left to the last. On the main issue which we have been discussing in this chapter—the rivalry between two methods of working a mechanical rule— what is the opinion of the pupils? After giving the two tests described above, I have never failed to ask the question: "Which of the two methods do you prefer? And the answer is always overwhelmingly in favour of the traditional method. Generally it is unanimous; always it is expressed with a gusto which leaves no doubt in the inquirer's mind. Whether I be right or wrong, the pupil is on my side.

CHAPTER XVI

LONG DIVISION

Le second (précepte), de diviser chacune des difficultés que j'examinerais en autant de parcelles qu'il se pourrait, et qu'il serait requis pour les mieux résoudre.

DESCARTES: *Discoura de la Methode.*

LONG division is supposed to be difficult—difficult to teach, difficult to learn, and difficult to work—so difficult in fact that we are sometimes advised to postpone teaching it till the pupil is eleven or twelve years of age. And since in the meantime he has to do a certain amount of dividing he has to make shift with short division. If the divisor is higher than twelve he must use factors. It is true that there are no factors of 13, or 17, or 19, or 23 or any other prime number; but we must avoid setting sums involving those divisors. It is true that when factors are used the remainder is perplexing, and the rule for finding it hard to remember; but the process, taken as a whole, is not so hard as long division anyhow. That is the line of reasoning adopted by the advocates of postponement. And I stoutly maintain that the reasoning is unsound.

The argument amounts to saying that the short-division road is the easier road. And this I flatly deny;

at any rate I deny that it is the easier in the long run. In the long run, taking the smooth with the rough, the pupil will find it a roundabout road, and a bewildering road, and a readily forgotten road, and a road beset with pitfalls—anything in fact but an easy road. Moreover, it is not the main road—not the king's highway, which must sooner or later be trod. And if the principles I have tried to expound are universally valid, as I think they are, to tread the king's highway from the very beginning is always the wisest plan.

It is wrong to regard long division and short division as two distinct rules. They are not. There is only one rule; and that is long division. Short division is long division with more of the work done in the head and less of it put down on paper. Or one may, with equal truth, put it the other way: there is only one rule, and that is short division, long division being short division set out in full.

In teaching long division we begin with what the pupils already know. They know *(a)*; we get them to proceed to *(b)*, and then to *(c)*:

$$
\begin{array}{lll}
& \underline{1837}\ \text{r. 2} & \underline{1837}\ \text{r. 2} \\
(a)\ 5\overline{)9187} & (b)\ 5\overline{)9187} & (c)\ 5\overline{)9187} \\
\quad 1837\ \text{r. 2} & & \qquad \underline{5} \\
& & \qquad 41 \\
& & \qquad \underline{40} \\
& & \qquad 18 \\
& & \qquad \underline{15} \\
& & \qquad 37 \\
& & \qquad \underline{35} \\
& & \qquad 2
\end{array}
$$

The step from *(a)* to *(b)* is simple: it is merely a matter of arrangement. The step from *(b)* to *(c)*, however, presents a difficulty. Let us take Descartes' advice and divide the difficulty. The long-division sum has a certain pattern, and the unit of pattern is this: $5\overline{)9}$. Let the

$$\frac{5}{4}$$

children learn this unit thoroughly. Let them set out $8 \div 3$, $19 \div 5$, $40 \div 8$, etc., in the same fashion, until they have completely mastered this fundamental step. Let them thus practise a double step, as in *(d)*, so as to learn how to treat the remainders. Other similar examples should be set. The children can readily check their answers in two ways: first, by short division, and secondly, by multiplication. They should then be set to work a number of examples, putting the short and long forms side by side as in *(b)* and *(c)* above.

(d)

$$
\begin{array}{r}
58 \text{ r. } 3 \\
4\overline{)235} \\
\underline{20} \\
35 \\
\underline{32} \\
3
\end{array}
$$

When the children are quite familiar with the pattern of the long-division sum they should be shown how to divide by such a number as 21, using both short division and long division as in *(e)* and *(f)*. In the earlier stages the multiplication table for the divisor should be completely written out before working the sum. This postpones the difficulty of finding a trial quotient, or, in other words, of guessing the next figure on the quotient, In the next stage the table is dispensed with and the pupils are put to divide by numbers of gradually increasing difficulty. The easiest divisors are 21, 31, 41, etc.

$$21 \times 1 = 21$$
$$2 = 42$$
$$3 = 63$$
$$4 = 84$$
$$5 = 105$$
$$6 = 126$$
$$7 = 147$$
$$8 = 168$$
$$9 = 189$$

```
              584 r.15              584 r.15
    (e) 21)12279       (f) 21)12279
                                   105
                                   177
                                   168
                                    99
                                    84
                                    15
```

The difficulty to be overcome at this point is that of finding the next figure in the answer. Suppose 85 has to be divided by 21 in one case, and by 29 in the other. In the first case we use 2 as the trial divisor and get 4 as the trial quotient. In the second case, as 29 is nearer 30 than 20, the best trial divisor is 3, not 2. The examples should be so graded that the children will get to realise that if a divisor lies between 50 and 60, 5 as a trial divisor will give the largest possible quotient, and 6 the smallest possible. The actual quotient will be one, or the other, or some intermediate number. My point is that finding the next figure in the quotient is a distinct difficulty which must be tackled by itself, and a good habit of procedure firmly inculcated upon the pupils.

There are other difficulties, such as appear when 15133 is divided by 37. The quotient is 409. There is always a danger of the 0 being omitted—a danger which is minimised when the quotient is placed *above* the dividend, and not, as in the olden days, to the right of the dividend. Here we have another example of a traditional method proving inferior to a new method.

Long division having been established as the standard rule-of-thumb method of dividing, no danger is incurred by dividing by "fancy methods." There is sometimes perhaps an advantage in dividing by factors, especially if there is no remainder, or if the exact remainder is of no consequence, or if decimals are used. For example:

$$378.5 \div 16 = \frac{378.5}{16} = \frac{94.625}{4} = 23.65625$$

This probably takes less time than if it is worked by long division. At any rate it took me 40 seconds to work it by the method given above and 45 seconds to work it by long division.

If the decimal does not terminate we can always get as accurate a result as is desired. In dealing with pounds sterling, for instance, it is rarely necessary to express the quantity to more than three places of decimals. Suppose, for instance, we wish to divide £438 16s. 10d. by 28, we may proceed, as in (g), to decimalise the money and divide successively by 4 and 7. I cannot show any

$$(g) \; \frac{£438 \; 16s. \; 10d.}{28} = \frac{£438.842}{28} = \frac{£109.711}{7}$$

$$= £15.673 = £15 \; 13s. \; 5\tfrac{1}{2}d.$$

enthusiasm for this method, since children take considerable time over it and are very inaccurate. It is certainly inferior to the standard method, which is exemplified by (h). This particular arrangement was invented by Professor Nunn.

(h)	£	s.	d.
	15	13	5½
28)438		16	10
28		360	144
158		376	154
140		28	140
18		96	14
		84	
		12	

When should our pupils learn the rule for dividing by factors? They should learn it when it is no longer necessary for them to learn it—after they have learnt a better method. I think they should be aware that such a method exists, that some people like it, and that they have themselves been using it in reduction. Suppose, for instance, it is required to reduce 1651 inches to yards. We may divide by 36, but the usual method is to divide successively by 12 and 3, which are factors of 36. Now compare the three examples: (*i*) where 1651 is divided by 36 (long division), (*k*) where 1651 inches are reduced to yards, and (*l*) where 1651 is divided by 36 by means of the factors 12 and 3.

```
            45
(i)  36)1651              (k) 12)1651 ins.
        144                    3)137 ft. 7 ins.
        211                       45 yds. 2 ft. 7 ins.
        180
         31
```

(*l*) 12)1651 ones

  3)137 twelves + 7 ones

   45 thirty-sixes + 2 twelves + 7 ones

Quotient, 45; Rem. : 2 × 12 + 7 = 31.

There are two important points to be borne in mind: first, that the remainder is of the same denomination as the dividend; second, that the quotient is of a different denomination—a different unit of value. Any examples that are worked should be set out in full as in example (*i*).

Divide 3893 by 45, using short division:

(*m*) 5)3893 ones

  9)778 fives + 3 ones

   86 forty-fives + 4 fives + 3 ones

Quotient, 86; Rem. 4 × 5 + 3 = 23.

CHAPTER XVII

EQUIVALENT FRACTIONS

And poise the cause in justices' equal scales.

SHAKESPEARE: *Henry VI, Part 2.*

IT is a simple and obvious fact that ½ = ²⁄₄; and it does not, at first blush, seem a very important fact. And yet, when expressed in the general form of $\frac{a}{b} = \frac{c}{d}$ it may be shown to embody principles which underlie a large number of our arithmetical rules and devices. Let me try to show how the equivalence of fractions may be made the starting-point of sundry mathematical excursions.

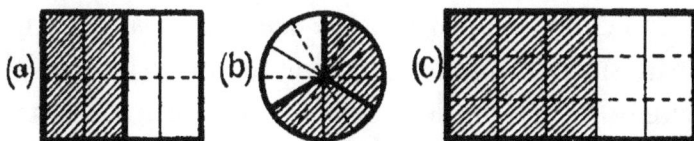

From Figure *(a)* it will be seen that ½ = ²⁄₄ = ⁴⁄₈
      "   "   *(b)*   "   "   ²⁄₃ = ⁴⁄₆ = ⁸⁄₁₂
      "   "   *(c)*   "   "   ³⁄₅ = ⁹⁄₁₅

Take any pair of these equal fractions, such as ²⁄₃ = ⁸⁄₁₂, and note the following facts about it:

THE ESSENTIALS OF ARITHMETIC

(1) 3 is just as many times as large as 2 as 12 is as large as 8.

(2) 8 is as many times as large as 2 as 12 is of 3.

(3) We may turn the fractions upside-down and they will still be equal. In other words the reciprocals of equals are equal.

(4) If we multiply crosswise we get equal numbers, i.e. $2 \times 12 = 3 \times 8$.

We can therefore from the original statement $\frac{2}{3} = \frac{8}{12}$ make the following inferences: (1) $\frac{3}{2} = \frac{12}{8}$, (2) $\frac{2}{8} = \frac{3}{12}$, (3) $\frac{8}{2} = \frac{12}{3}$, (4) $2 \times 12 = 3 \times 8$.

If we try any other pairs of equivalent fractions we shall find that the same inferences may legitimately be made. In fact, quite young children can study the principles of proportion from simple diagrams and fractions.

When the pupils are familiar with the rules for the multiplication and division of vulgar fractions they may be invited to approach the equivalence of fractions from another angle. We will assume that the rule for multiplication is known in the usual form, and the rule for division in this form: Divide the numerator of the dividend by the numerator of the divisor to obtain the new numerator, and divide the denominator of the dividend by the denominator of the divisor to obtain the new denominator. To put it algebraically:

$$\frac{a}{b} \div \frac{c}{d} = \frac{a \div c}{b \div d}$$

Now submit the question: What number is that

which will yield the same result whether it is used as a multiplier or a divisor? The answer, of course, is 1. If we multiply or divide any number of 1 the number remains unchanged. But $1 = \frac{2}{2} = \frac{3}{3} = \frac{4}{4}$, etc. $\frac{1}{2} \times 1 = \frac{1}{2}$, $\therefore$ $\frac{1}{2} \times \frac{2}{2} = \frac{1}{2}$, $\therefore \frac{2}{4} = \frac{1}{2}$. Again, $\frac{2}{4} \div 1 = \frac{2}{4}$, $\therefore \frac{2}{4} \div \frac{2}{2} = \frac{2}{4}$, $\therefore 2 \div \frac{2}{4} \div 2 = \frac{2}{4}$, $\therefore \frac{1}{2} = \frac{2}{4}$.

From these facts two important laws emerge. (1) If we divide the numerator and denominator of a fraction by the same number, the value of the fraction remains unchanged. (2) If we multiply the numerator and denominator of a fraction by the same number, the value of the fraction remains unchanged. It is important to note that we neither multiply nor divide the fraction by the number in question: in each case we either multiply by 1 or divide by 1.

This is the logical basis of all cancelling. We may cancel by dividing by 1 (in this instance by $\frac{7}{7}$), as when we say that $\frac{\overset{2}{\cancel{14}}}{\underset{3}{\cancel{21}}} = \frac{2}{3}$; or we may cancel by multiplying by 1 (in this instance by $\frac{100}{100}$), as when we say that $\frac{\overset{3}{\cancel{.03}}}{\underset{40}{.4}} = \frac{3}{40}$; or (by $\frac{4}{4}$ this time) that $\frac{\overset{12}{\cancel{3}}}{\underset{100}{\cancel{25}}} = .12$.

A quantity is usually simplified by the former kind of cancelling (by division); but the latter kind is of special service when we have to deal with a complex fraction involving decimals, such as $\dfrac{.06 \times 3.14 \times 2.7}{1.5 \times 12.65 \times .003}$.

When multiplied by $\dfrac{10^6}{10^6}$ the expression becomes

$\dfrac{60 \times 314 \times 27}{15 \times 1265 \times 003}$, and this may be simplified by the commoner form of cancelling.

Division by factors is based on the same principle (see Chapter XVI). For example, $14.26 \div 2.1 = \dfrac{14.26}{2.1} =$

$\dfrac{14.26}{2.1} \times \dfrac{10}{10} = \dfrac{142.6}{21} = \dfrac{142.6}{21} \div \dfrac{3}{3} = \dfrac{47.533}{7} = 6.790$, correct to three places of decimals.

Let us now return to simple equivalent fractions and consider how we may find a missing term. If $\dfrac{2}{3} = \dfrac{x}{6}$ is quite easy to find x by mere inspection; but if $\dfrac{2}{3} = \dfrac{x}{10}$ the value of x is not so obvious. It can, however, readily be discovered by making use of the fact that $3x = 2 \times 10$. For x is now seen to be equal to $\dfrac{20}{3}$, or $6\dfrac{2}{3}$, or 6.667 approximately. Here we find a starting-point for three important lines of study: the laws of proportion, the theory of the equation, and the interpretation and manipulation of formulæ.

In no practical sphere is the principle of the equivalence of fractions more useful than in the division of decimals. Suppose, for instance, we have to divide 14.375 by 0.17. The first step is to arrange the figures in the form of a fraction, thus: $\dfrac{14.375}{0.17}$. If we originally fix the decimal points directly underneath each other we can legitimately move these points (so long as we move them both together) to the right or to the left to any position indicated by a vertical line passing between the figures. Thus, $\dfrac{14.375}{0.17} = \dfrac{0.14375}{0.0017} = \dfrac{1.4375}{0.017} = \dfrac{143.75}{1.7} =$

$\dfrac{1437.5}{17} = \dfrac{14375}{170}$. Each of three rules in vogue for the division of decimals depends on first writing the fraction in one or other of the three last forms given above. If the standard-form method is adopted, the divisor has to be reduced to 1.7; the method I advocate is to reduce the divisor to 17, that is, to a whole number; the method I learnt as a boy was to reduce both dividend and divisor to whole numbers, as in the last form above.

The advantage of the standard-form method is that an approximate answer may readily be obtained by a very simple calculation. It is obvious from looking at the standard form of the above example that the answer lies somewhere between 72 and 144. This is not a very close approximation it is true, but it is close enough to keep the decimal point of the quotient in its right position. I am of opinion that the best rule-of-thumb method of dividing decimals follows the simple maxim: Make the divisor a whole number. I should, for instance, work the above example like this: $14.375 \div 0.17 = \dfrac{14.375}{0.17}$

(*Imagining* the points moved one step to the right so that the divisor is in standard form, I should note that the answer should come between 72 and 144, but nearer the former than the latter.)

$$\frac{14.375}{0.17} = \frac{1437.5}{17} = \frac{84.559}{17)1437.5}$$

Answer: 84.559, true to three places.

There is not the same objection to the "standard-form" method in division as in multiplication; for the rule for division, being in accordance with the principles of cancelling, is quite easy to remember. It is only when the "standard-form" method is used for multiplication as well that there is liability to confusion—a confusion which invades both operations. Those who have tried the standard-form method for both and have then discarded it in multiplication, but retained it in division, have found that the results in both multiplication and division have markedly improved.

If the method of limits be adopted, there is a "snag" to be warned against. Suppose, for instance, we have to divide 3115.4 by 36.25. $\frac{3115.4}{36.25} = \frac{311.54}{3.625}$. The highest value for this is not $\frac{312}{4}$, nor is the lowest $\frac{311}{3}$. If it were, we should be in the absurd position of having the lowest value higher than the highest. The highest value is secured by taking the higher value of the numerator and the lower value of the denominator; and vice versa for the lowest value. In other words the fractions should be $\frac{312}{3}$ and $\frac{311}{4}$ and respectively.

The main purpose of this chapter has been to show that the principle of the equivalence of fractions pervades a wide tract of the arithmetical field. It appears in cancelling, in the reduction of a fraction to its lowest terms, in the simplification of fractional forms, in rule of three, in simple equations, in the manipulation of formulæ, and in the division of decimals. It is important, therefore, that our pupils should have a firm grasp of the principle.

CHAPTER XVIII

VULGAR AND
DECIMAL FRACTIONS

The fragments, scraps, and bits.

SHAKESPEARE: *Troilus and Cressida.*

"Madam," I said, "you can pour three gills and three quarters of honey from the pint jug, if it is full, in less than one minute; but, Madam, you could not empty that last quarter of a gill, though you were turned into a marble Hobe, and held the vessel upside down for a thousand years."

O. W. HOLMES: *The Autocrat of the Breakfast Table.*

WHAT happens when we measure the length of a room? We take a fairly large unit such as a foot and find how many times a footrule can be laid along the room. We have now measured the main distance. There is a piece over; so we take a smaller unit, one twelfth as large, and measure the excess with this smaller unit. There is still a little left over. So we take a still smaller unit— say, half an inch or a quarter of an inch. If we record the length as 15 feet 3¼ inches we shall be regarded as scrupulously, and perhaps unnecessarily, exact. It is the 15 feet that is important. The rest is comparatively trivial.

So in weighing. We use the heaviest weight first and find the gross and substance of the thing. Then we weigh the small excess in ever-diminishing detail. When the larger unit becomes too coarse, we have to resort to a finer one. Thus we record weight in hundredweights, quarters, and pounds, time in hours, minutes, and seconds, or capacity in gallons, quarts, and pints. We use units of a descending order of magnitude. If we have no smaller units that are conventionalised, we invent some for ourselves and call them fractions. We have in the English system no unit of weight less than an ounce, nor of length less than an inch. So when the ounce or the inch become too coarse for our purpose we use half-ounces, or quarter inches or some other fractional parts. There is in all this a sad lack of system and of orderliness. The result is expressed in figures whose values descend in steps, but the steps are extremely irregular. From the value of one unit we can never infer the value of the next: each has to have its own identification label.

When we express a quantity by a mixed number, as in "$4\frac{2}{3}$ hours," we are using two units—one a standard unit and the other an arbitrary subdivision of that unit. But even this liberty of choice in the smaller unit does not bring us nearer to an ideal system. Suppose I have a piece of metal about $2\frac{1}{2}$ inches long and I wish to measure it as accurately as possible. By means of a foot-rule marked in twelfths I might estimate it at $2\frac{5}{12}$ inches. Another rule marked in sixteenths might give me $2\frac{7}{16}$ inches. But however finely graded the inch may be, I can never discover the exact length of my piece of metal. This question is more fully discussed in the chapter on Incommensurables; what concerns us here is that a fraction with a large denominator does not secure its avowed purpose. And besides, it is terribly difficult to deal with afterwards. But then nobody as a matter of fact measures this way. If he aims at a high degree of accuracy he uses decimals, and proceeds by a method of diminishing steps. With a decimal scale he estimates the length as, say, 2.42 inches (he would need a micrometer to get the second place of decimals), using three units—inches, tenths of an inch, and hundredths of an inch—related to one another in definite orderly fashion. This method enables him to estimate the length to the degree of precision that suits his purpose. Roughly, the piece of metal is 2 inches long; more precisely it is 2.4 inches long; more precisely still it is 2.42 inches long. Munition workers during the war were sometimes required to carry the figures to four places of decimals, and get their measurements right to within the ten-thousandth part of an inch.

Thus a decimal fraction is expressed in a series of digits of regularly diminishing value. It resembles a diminuendo passage in music, except that each note gets suddenly fainter than its predecessor. After the third or fourth note the music becomes almost inaudible, and it is no longer profitable to listen.

Simple vulgar fractions are easier to understand than decimal fractions; but once the full import of the decimal system is understood, decimals convey a more useful notion of magnitude and are of much greater practical value. Dr. G. M. Wilson recently made a survey of the social and business usage of arithmetic in America. He discovered that over 95 percent of the fractions actually used were included in this list, ½, ¼, ¾, ⅓, ⅔, ⅛, ⅜, ⅕, ⅖, and ⅘. Now look at this example extracted from a textbook which is extensively used in English secondary schools:

Find the value of $\dfrac{11}{17} + \dfrac{31}{51} + \dfrac{267}{357} + \dfrac{5}{13} + \dfrac{24}{39}$.

Take the simplest of these fractions, ⁵⁄₁₃. In what conceivable circumstances could a person arrive at such a fraction? Certainly not by weighing or measuring. No scale of weights or measures is ever graduated in thirteenths. Does it suggest the way one might serve five soles to thirteen persons seated at table? Well, suppose we secure our fraction ⁵⁄₁₃, what are we going to do with it? Is it anything more than a curiosity? Who wants to add ⁵⁄₁₃ to anything else, anyhow? It may, it is true, appear as a ratio in a proportion sum; but although we may need to multiply or divide this ratio, we shall never need to add it or subtract it. If we did, the simplest plan would be to reduce it to a decimal. An example such as I have just quoted aims at giving practice in finding the L.C.M. But in practical life there is little need for finding the L.C.M. If we have to add fractions, they are generally measurements of some kind—and of the same kind. Halves and quarters and eighths and sixteenths go together. We never find them mixed up with thirteenths and seventeenths and thirty-ninths. The highest denominator is generally the least common multiple of the lot. If not, we can double it, and see if it is then common. Even if we overshoot the mark and take a common multiple which is not the least common multiple, it doesn't matter. It only means that the resulting fraction is not in its lowest terms.

Pupils in the olden days were frequently required to arrange a series of fractions in order of magnitude. They are sometimes required to do so now. The standard rule was to reduce them to a common denominator and compare their numerators. There was a subsidiary rule: Reduce them to a common numerator and compare their denominators. There is another rule, much better than either of them: Reduce to decimals and then compare. Here is an example:

Arrange the following fractions in descending order of magnitude:

$$\frac{67}{92}, \frac{54}{85}, \frac{56}{79}, \frac{196}{237}.$$

To judge their relative values by inspection is an impossible task. To compare them by reducing either to a common denominator or to a common numerator is an interminable task. Now try decimalising them. First reduce them to one place only. They become respectively .7 . . . , .6 . . . , .7 . . . , and .8 . . . . It is immediately clear that the last is the biggest and the second the smallest We have now to decide between the first and the third. Carrying the conversion one place further we find the first to be .72 . . . and the third .70 . . . . That settles the whole list.

If we had to add those fractions we should reduce them to decimals true to three or four places and then add them. It would save much trouble and be accurate enough for any reasonable purpose.

My little daughter was much puzzled one evening over a homework sum. She had multiplied by a fraction and had got a result which was less than the original number. She argued that it was impossible to make a thing smaller by multiplying, and yet this had actually happened when she had multiplied by ⅔ according to rule. The next morning, by a curious coincidence, a colleague told me a similar story about his own little girl. As a matter of fact we have here a difficulty which all children sooner or later encounter. It is long before they get over their surprise at the topsy-turvydom caused by a fractional multiplier or divisor. Ask a class of pupils of 14 or 15 years of age to divide 8 by ½ and a considerable number will almost certainly say 4.

Examples such as these should be studied by the children:

$8 \times 2 = 16$	$8 \times 2 = 16$	$8 \div 2 = 4$	$8 \div 2 = 4$
$8 \times 1 = 8$	$8 \times 1 = 8$	$8 \div 1 = 8$	$8 \div 1 = 8$
$8 \times \frac{1}{2} = 4$	$8 \times .4 = 3.2$	$8 \div \frac{1}{2} = 16$	$8 \div .4 = 20$
$8 \times \frac{1}{4} = 2$	$8 \times .2 = 1.6$	$8 \div \frac{1}{4} = 32$	$8 \div .2 = 40$
$8 \times \frac{1}{8} = 1$	$8 \times .1 = .8$	$8 \div \frac{1}{8} = 64$	$8 \div .1 = 80$
$8 \times \frac{1}{16} = \frac{1}{2}$	$8 \times .01 = .08$	$8 \div \frac{1}{16} = 128$	$8 \div .01 = 800$

The pupils should then be able to formulate for themselves some such rule as this. When a number is multiplied by a number greater than 1, it becomes greater; when multiplied by 1 it remains the same; where multiplied by a fraction it becomes less. A corresponding rule may be deduced for division.

We will now consider the decimalisation of money. Pupils used to be asked such a question as this: What decimal of 14s. 7¾d. is 2s. 10½d.? Such a question is seldom asked now. There is no point in asking it. Nobody ever wants to know. There is, however, much to be said for reducing money to the decimal of a pound. For although we have not yet adopted a decimal coinage, it is often easier to calculate when the money is in the form of decimals than when it is in the form of £ s. d. It is certainly easier in finding compound interest. So practice in decimalising English money is desirable. The old way was to find what fraction of one pound the given sum of money was, and then convert the vulgar fraction into a decimal fraction. But there are two better ways than that. The first way is to begin decimalising at the lower end, dividing the pence by 12 and the shillings by 20. I give two examples:

Express as decimals of £1 (a) £3 12s. 4½d., (b) £5 13s. 8½d.

(a)				(b)		
£	s.	d.		£	s.	d.
3	12	4½		5	13	8½
= 3	12	4.5		= 5	13	8.5
= 3	12.375			= 5	13.708333 . . .	
= 3.61875				= 5.68541666 . . .		
= 3.619 approx.				= 5.685 approx.		

The second way is based upon the fact that 1 florin = £.1, and 1 farthing = £.001¼ . In example (a) the 12s. = £.6, and the 18 farthings = £.018¹⁸⁄₂₄ = £.018¾ = £.01875. The whole sum therefore equals £3.61875.

In example *(b)* the 13*s.* = £.65, and the 34 farthings = £.0344³⁴⁄₂₄ = £.035¹⁰⁄₂₄ = £.035⁵⁄₁₂ = £.03541666 . . . The complete answer is £5.68541666 . . .

The second method is very simple when it is required to express a sum of money to three decimal places only. Roughly speaking, £.1 is put for every florin and £.001 for every farthing besides. This makes the decimal right to the nearest penny. If £.000 ^{x}⁄₂₄ is added for x farthings, it makes it right exactly. If the nearest whole to ^{x}⁄₂₄ is added, the answer is right to the nearest farthing.

There is a corresponding rule for converting decimal money into £ *s. d.*, which depends upon the fact that £.1 = 1 florin and £.001 = 1 farthing − ½₅ farthing. On the whole the simplest plan is to multiply the fractional parts successively (in one's head) by 20 and by 12. There is no need to multiply by 4, as it is seen at a glance how many farthings there are. For example: Change £3.61875 to £ *s. d.*

> £3.61875
> 12.375*s.*
> 4.5*d.*   *Answer* £3 12*s.* 4½*d.*

It will be observed that the sum of money in *(a)* can be exactly expressed as a decimal, while the sum in *(b)* cannot. Is there a rule by which we may tell at a glance whether a sum can or cannot? There is. All shillings will decimalise exactly, so we need not trouble about them. Convert the rest—the pence and farthings—into farthings. If the number is a multiple of three the whole sum of money will decimalise exactly. If not, it won't. The reason will be found in the next chapter.

CHAPTER XIX

RECURRING DECIMALS

Thou hast damnable iteration.

SHAKESPEARE: *King Henry IV, Part I.*

RECURRING decimals have quite gone out of fashion. There was a time when they had a great vogue—when page after page of the textbook was filled with rules and examples in recurring decimals. Now they have quite disappeared from the textbook and the course of study. Examiners have ceased to set questions on them, and the syllabus of the First School Examinations expressly excludes them. The upshot is that the modern school child never hears about recurring decimals: he is not supposed to be aware that they exist.

Why this conspiracy of silence? If we are to find decimal equivalents for vulgar fractions, a large number of them, indeed an overwhelming majority of them, must be recurrers. If we leave out the simplest of the vulgar fractions—the halves, quarters, and eighths—the odds are heavily against a vulgar fraction producing a terminating decimal. If a fraction is in its lowest terms and its denominator contains any other prime factor besides 2 and 5, it must generate a repeating decimal.

For instance,

$$\frac{3}{4} = \frac{3}{2 \times 2} = \frac{3 \times 5 \times 5}{(2 \times 5) \times (2 \times 5)} = \frac{75}{10 \times 10} = .75$$

This terminates. But no such device will make ⅔ terminate. By multiplying a fraction by 2/2 for every 5 as a factor in the denominator, or by 5/5 for every 2, we can always obtain a fraction whose denominator is ten or a power of ten. That is, we can always express it exactly and completely as a decimal fraction. But if a prime factor other than 2 and 5 intrudes in the denominator, no integral factor can turn the intruder into a power of ten, and the corresponding decimal fraction cannot terminate. So no thirds, or sixths, or sevenths, or ninths, or elevenths, etc., if they are in their lowest terms, can find exact equivalents in the decimal system.

It is now clear why a sum of money will decimalise exactly if the number of farthings is a multiple of 3. Expressed as a vulgar fraction it cannot have more prime factors in the denominator than those of 960, the number of farthings in a pound. But 960, having as prime factors $2 \times 2 \times 2 \times 2 \times 2 \times 2 \times 3 \times 5$, contains but one that is refractory, namely 3. When this is removed by cancellation, there are none left but 2 and 5.

The young arithmetician cannot go far before he comes upon the strange phenomenon of a division sum which never ends. He sees that ⅓ = .3333 . . . *ad infinitum,* and that 3/11 = .272727 . . . *ad infinitum.* He is asked to decimalise money and finds that he can't always do it exactly. And it is surely of interest, and of some slight importance, to know beforehand whether the sum of money can or cannot be exactly expressed as a decimal. Especially if it has to be multiplied. For to multiply an inexact quantity is to multiply its error as well. If the sum of money is wrong by a quarter of a farthing to begin with, and it has to be multiplied by 10,000, it is ultimately wrong by over £2 12*s*.

My contention is that our pupils should have at least a nodding acquaintance with recurring decimals. They should be aware of their frequency and their significance. They should be able to discover without actually dividing out whether a vulgar fraction will or will not produce a recurring decimal. They should realise that a recurring decimal is not an irrational number—that although the figures are supposed to go on for ever, the magnitude they represent is quite definite and is indeed capable of being exactly expressed as a vulgar fraction. They should be acquainted with the symbolism by which recurrence is expressed, and should know in its simplest form the rule for changing a recurrer back into a vulgar fraction.

If the children should not know all these things, the teacher at any rate should. It has been possible of recent years for teachers to leave the training college and take general charge of a class in an elementary school without knowing any arithmetic beyond the matriculation standard. These would be well-advised to devote a little time to the study of terminating and recurring decimals.

A recurring decimal is an endless series of terms in geometrical progression. $.3 = 3r + 3r^2 + 3r^3$, etc., where $r = \frac{1}{10}$. It is quite impossible, therefore, to deal adequately with the topic until this type of series has been studied. It is, however, possible to make an inductive study of recurrers and arrive at simple empirical rules. Since $\frac{1}{9}$ = .111 . . . = .1, and 2/9 = .222 . . . = .2, and $\frac{26}{99}$ = .2626... = .26, and $\frac{145}{999}$ = .145145... = .145, we may provisionally infer the rule for converting a recurring decimal to a vulgar fraction. Place a nine in the denominator for every recurring figure. But how are we to treat mixed recurrers? The traditional rule is complex and difficult to remember. There is, in fact, no need to remember it—no need for any rule other than the simple one given above. This rule may be extended so as to cover all cases. We can, for instance, change .63 into a vulgar fraction by substituting .6 $\frac{3}{9}$, which equals $\frac{6\frac{1}{3}}{10}$, or $\frac{19}{30.}$.

Recurrers have one great advantage: they are interesting. One might almost call them comic. One-third producing an endless series of threes suggests a conjurer producing an endless ribbon out of his mouth. .9 turning out to be nothing but our old friend 1 has all the magic of a transformation scene at a pantomime.

The dull soul to whom the comic side of recurrers does not appeal need not convert them to vulgar fractions at all. He can treat them as we usually treat decimals that stretch to an unprofitable length—he can cut them short. It is sufficient for all practical purposes to regard one-third as .333, and two-thirds as .667. It is just as easy to add and subtract them thus as it is to add and subtract vulgar fractions. True, it is not so easy to multiply and divide them. But then we cannot afford to despise even such things as recurring decimals without paying some sort of penalty.

CHAPTER XX

PROBLEMS

She came to prove him with hard questions.

1 Kings x. 1.

THE teaching of problems is itself a problem. The teacher of the last generation solved it quite simply by not teaching them at all. He argued that it was his business to give his pupils arithmetic, not to give them brains. He taught them the rules, and if they had the brains they would apply the rules; if they had not—well, that was no fault of his. There was in this tenet a certain measure of common sense. As a working hypothesis it stood its ground fairly well. The results were not brilliant, it is true; but neither were they contemptible: they were just middling.

Then came a body of reformers, proclaiming far and wide the gospel of Intelligence, and calling sinners to repentance. The sinners were those who held the heresy that it was not their business to cultivate brains. The true faith was that brains were the only things worth cultivating. Everything else was dross in comparison. Whether a pupil had more or less arithmetic when he left school was not of much consequence: what really mattered was the amount of Intelligence he had acquired. And problems in arithmetic were the very things to give him this Intelligence, and give it him in abundance. They stimulated the reasoning powers. They sharpened the wits, they quickened the understanding. They turned the dull into the sensible, the sensible into the brilliant, and the brilliant they made shine with inconceivable splendour. This was the theory, this the promise; but the fulfilment never came. There was somewhere a flaw—a very serious flaw—in the reasoning. In the first chapter I tried to indicate the nature of the flaw in the theory in general; in this chapter I will try to show the fallacy that lurks in the theory in particular—the theory as it applies to problems.

We cannot recognise an example as a problem by merely looking at it. The evidence is incomplete. We must know something about the mind that has to deal with it. For the distinguishing mark of a problem is novelty; and novelty is a relative term. What is new to one child is not new to another. Even 3 + 2 is a problem when encountered for the first time; and to a practised mathematician the most ghastly complication of symbols and quantities may be a mere routine sum. It depends on what the mind is already familiar with. In a problem a new step has to be taken; and the size of the step is determined by the mind's native intelligence. An able child can take a big step; a stupid child only a little one. The bigness, however, is only comparative. Absolutely, both steps are small.

Let me illustrate by an example. Forty or fifty years ago the mathematical lecturer at the old Borough Road

Training College was a man of much originality. Though "Fate tried to conceal him by naming him Smith," he had acquired no small measure of fame through his scheme of geometry. When a new batch of students entered the college his first step was to convince them that they knew nothing; that they had not even mastered the first book of Euclid. To prove his charge he set them an examination which consisted of three "deductions." In the first, they were required to prove that the lines bisecting the three angles of a triangle met at a point; in the second, that the lines drawn at right angles to the sides of a triangle from their mid-points met at a point; in the third, that the perpendicular dropped from the angles to the opposite sides met in a point. These are simple problems in concurrence with which, since they now appear in all textbooks, the modern student is well familiar. But they were to be found in no textbook in the days of which I write; and to the students in question they were wholly new and strange. The difficulty was to prove that *three* lines met at a point. To prove that two lines met was quite easy; but how was it possible to prove that a third line would pass through the very same point as the other two? The result of the test was invariably the same: all the students would be floored. Mr. Smith, considering them to be now in a proper frame of mind to receive instruction, would proceed to solve the first problem. That was enough. The other two the students solved for themselves without very much difficulty. It was the first step that counted—the step that separated this new type of "rider" from all the others they had ever met. It was a step beyond their powers. The other two steps were small in comparison.

For the general pattern of the reasoning was the same for all three theorems: it was in the details alone that they differed.

It will thus be seen that when the new step which the mind is invited to take is beyond its capacity, the best expedient is to shorten the step. In the example given above the leap which landed the class in the new territory was taken by the tutor. The class was simply carried there. It would have been possible, however, by means of a series of intermediary exercises, to induce the students to bridge the gap for themselves. But it would have meant taking two or three strides instead of one. The difficulty is met, not by stretching the mind, but by shortening the stride.

To put it in another way, a pupil's capacity to work a given problem depends partly on his native intelligence (for which the teacher is not responsible) and partly on his familiarity with similar instances (for which the teacher is responsible). Hence the only thing a teacher can do is to teach as many types as possible. It is of small avail to let the pupil loose among a multitude of new and miscellaneous problems in the hope that by sheer force of intellect he will solve them for himself. He will simply miss his way and lose his courage. To secure progress in arithmetic we must organise the material, we must classify and cross-classify, we must help our pupils to discern an underlying similarity of pattern in a large variety of examples. We must, in fact, reduce our examples to types, and teach the types. The commoner the type, the more important it is to teach it. Simple addition, simple multiplication, and so forth, are of such wide applicability that nobody has doubted the wisdom of fixing a standard procedure. If justification were needed for teaching the "rules" in arithmetic it would be found in the fact that a large number of sums fall into the same pattern, and when the pattern is known, the difficulties due to change of material can be coped with by all but the dullest. But though nobody has doubted the wisdom of revealing the way in which the commoner sums fall into distinct types or patterns, many people have doubted the wisdom of treating problems in this manner. Problems are supposed by them to belong to a province of their own—a land of anarchy and confusion, where every member is a law unto himself. But as I have already shown, there is no clear line of demarcation between the mechanical sum and the problem; and there need be no point at which we must cease classifying. We classify as far as it is expedient, and no farther.

Let us consider these two examples:

(a) Share 3*s*. 10*d*. equally between two boys.

(b) Share 3*s*. 10*d*. between two boys so that one has 6*d*. more than the other.

Most teachers would pronounce the first a routine sum, and the second a problem. The first is a simple sum in division of money. The rule having been definitely taught, the precise mode of procedure is familiar to the pupil. Although he may never have actually divided 3*s*. 10*d*. by 2 in his life before, he has done something so like it that the operation presents to him no difficulty. There is nothing new in the pattern: there is no problem.

The second example has so frequently appeared in recent arithmetic tests that it has come to be regarded as a definite type. To the average pupil it has become a routine sum. He is familiar with the rites to be observed; he has seen a sample worked on the blackboard, and has worked others on the same model. There is nothing newer to him in the second example than in the first, and the second is just as much a mechanical sum as the first.

Let us suppose, however, that the pupil has encountered this kind of sum for the first time. It would then be a genuine problem. To me, at the mature age of twenty, although I had read a fair amount of mathematics, it was a genuine problem. I then met it for the first time; not in a classroom, but in a restaurant. I paid a bill for two, and my companion had to settle with me afterwards. I forget the exact amounts, but we will assume them to be the same as in example *(b)* above. My companion had to pay 6*d.* more than I. I well remember that we were both a little puzzled how to proceed. After a moment's thought I suggested that we should halve the whole bill, halve the difference between our shares, and add the half difference to make the larger share, and subtract it to make the smaller. Thus my bill would come to 1*s.* 11*d.* — 3*d.*, and my friend's to 1*s.* 11*d.* + 3*d.* The method was clumsy, but the reasoning was sound and the solution was correct. Children are taught nowadays to put the difference (6*d.*) aside and divide the remainder. Half the remainder gives one share, and half the remainder plus 6d. gives the other. They are taught to take these steps in regular order as a fixed piece of procedure. The self-same example which a generation ago was a problem to a youth of twenty is nowadays a routine sum to a child of ten.

This is an interesting little problem, interesting because the same basic pattern runs through a variety of examples. So varied are they that there is no small difficulty in identifying the pattern. Note these instances:

(1) Find two numbers whose sum is 28 and whose difference is 6.

(2) A bottle and a cork cost 2½*d.* The bottle cost 2*d.* more than the cork. What was the price of the cork?

(3) A man rows down the stream at the rate of 6 miles an hour, and against the stream at the rate of 3 miles an hour. What is the rate of the stream?

These, then, are essentially of the same type as example *(b)* above.

Having given an instance of a type of problem which has been dragged from the obscurity of the rabble and given recognition as a definite class, I will now give an instance of a type that is in danger of sinking back into the great unclassed—of becoming *démodé* and losing its label—of sharing, in fact, the fate of barter, tare and tret, and chain rule. The type in question used to be called Alligation and given a chapter to itself in the text-books. It was regarded as a monopoly of the tea-mixer, and generally took some such form as this: In what proportion must a grocer mix tea at 3*s.* a lb. with tea at 2*s.* a lb. in order to be able to sell it at 2*s.* 7*d.* a lb.? If it merely applied to the blending of tea, its passing away need cause no grief. But it has a much wider application, and the principle which underlies it is well worth our study. It is really a problem in averages. Having been given an average, and a clue to the items on which the average is based, we are asked to find the items themselves. In fact we have to work an "average" sum backwards. In the example here worked in full:

> 7 lb. at 3*s*. per lb. = 21*s*.
> 5 lb. at 2*s*. per lb. = 10*s*.
> 12 lb. cost = 31*s*.
> Average cost per lb. = 2*s*. 7*d*.

we are given the average, 2*s*. 7*d*., part of the first item, 3*s*., and part of the second item, 2*s*.; and we are required to find the quantities 7 lb. and 5 lb.

Let us experiment with these figures. Double both the quantities; the average remains unchanged. So it is the ratio between the two quantities that matters, not the actual quantities. Next, it is clear that the average must always be somewhere between the 3*s*. and the 2*s*. Make the quantities equal and the average price is midway between the two original prices. Make the ratio of the quantities 2 to 1, and the average price becomes 2*s*. 8*d*.; make it 1 to 2 and the average is 2*s*. 4*d*. Change the ratios to 3 to 1 and to 1 to 3, and the average prices become 2*s*. 9*d*. and 2*s*. 3*d*. respectively. From these simple experiments we draw the following conclusions:

(1) It is the ratio of the quantities and not the absolute quantities that determines the average price.

(2) The larger quantity drags the average price towards its own side.

(3) The distances of the average from the two original prices are inversely proportional to the quantities.

The last conclusion is not so obvious as the others and needs further illustration. This is afforded in the following diagram:

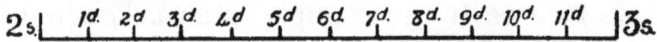

If 2s. xd. represents the average price of the mixture, the ratio of the quantities is given in this table:

When *x* is l*d*., the ratio is 11 of the cheaper to 1 of the dearer.

When *x* is 2*d*., the ratio is 10 of the cheaper to 2 of the dearer.

When *x* is 3*d*., the ratio is 9 of the cheaper to 3 of the dearer, etc.

Hence this rule for making the original example, the numbers representing pence:

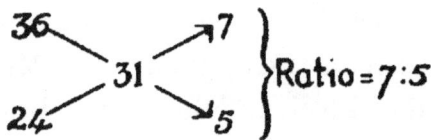

The usefulness of this rule is not limited to the mixing of tea. Many years ago an old pupil of mine who was at a training college (a women's training college) sent me a problem which was alleged to have baffled the whole staff. The exact figures have escaped my memory, but the problem ran something like this: "I invest £1092 partly in the 3 percents at 84 and partly in the 4 percents at 96. My entire income from these investments is £43. How much money did I invest in each fund?" The failure to solve this problem was due to the fact that the rule for alligation was at one time out of fashion. It seems to be coming back again. At any rate, a fair number of the questions recently set at the various examinations for the First School Certificate are of the alligation type.

It is sometimes urged that many, if not most, of the problems set in arithmetic should be solved by algebra. With this I quite agree; especially if the words "as well" are added to the last sentence. I think there is something gained by an arithmetical analysis which is missed by the algebraic solution. Let us try the alligation problem given above:

Let x = the number of lb. at 36d. a lb.
and y = "   "   " " " 24d. " "

Then $36x + 24y = 31(x + y)$
$$36x + 24y = 31x + 31y$$
$$5x = 7y$$
$$\frac{x}{y} = \frac{7}{5}$$

This is an unusual type of equation—a type in which the absolute values of x and y are not sought for (indeed cannot be found), but simply the ratio between them. And it is very doubtful whether a pupil could evolve such a solution from his knowledge of the commonplace equation. The type would have to be specifically taught. Moreover, it seems to me that the full implications of the problem are more clearly brought out in the arithmetical solution than in the algebraic.

Let me close the chapter by stating clearly what my general thesis is, and what it is not. I contend that there is no way by which we can teach problems in general: all we can do is to teach special kinds of problems. We have to show that some of them are akin and fall into groups. We have to exhibit the pattern that is common to the group and the line of reasoning by which solution is reached. I do not mean that each problem is to be presented with a label attached; I do not mean that problems of the same class should be put together for purposes of practice; I do not mean that a textbook should attempt a full classification of problems and try to establish each type as a separate rule. The classification I advocate is for the teacher, not the pupil. When presenting a problem the teacher himself should know quite clearly whether it involves principles or procedures which are of wide applicability. Problems should, as a rule, be given the pupils without a word of explanation. They should be attacked by what seem to them to be commonsense methods. The mind should have a sense of elbow-room and complete freedom of attack. But here lies the point: the success of the attack will depend upon a subconscious identification of the type; and this subconscious identification of the type will depend upon the skill with which the teacher has prepared the ground.

CHAPTER XXI

PROBLEMS UNDER CRITICISM

> Problems are short stories of adventure and industry with the end omitted.
>
> STEPHEN LEACOCK: *Literary Lapses.*

IF we try to follow the negative advice given in various books on diet, we shall die of starvation. The total mass of prohibitions leave us nothing to eat. We are placed in much the same quandary by the critics of problems in arithmetic. We are told by one to avoid this, by another to avoid that. Discard all the problems declared to be bad and there are none left to teach. It sometimes happens that one writer turns the advice of another writer upside-down. Most writers condemn clock sums and commend a free use of British weights and measures. Sir Oliver Lodge condemns British weights and measures as antiquated and needlessly intricate,[19] and himself works clock sums as interesting examples which make a demand upon the pupils' capacity to think.[20] It must be pointed out, however, that he uses algebraic symbols in solving them.

[19] *Easy Mathematics,* p. 63.

[20] *Idem,* p. 110.

Thorndike would exclude all problems whose answers would in real life be already known. He gives as an example:

"A clerk in an office addressed letters according to a given list. After she had addressed 2500, ⁴/₉ of the names on the list had not been used; how many names were in the entire list?"

This principle, if adopted, would cut out at a stroke a huge number of problems from our textbook and examination papers. It prohibits such examples as: "I am thinking of a number. Half of this number is twice six. What is the number?" He admits, however, that this is better than the following because it makes no false pretences: "A man left his wife a certain sum of money. Half of what he left her was twice as much as he left to his son, who receives $6000. How much did he leave his wife?" [21]

We have at various times been counselled to omit from the arithmetic programme all problems which are not interesting to children; which do not serve the needs of child life; which do not serve the needs of social and business life; which have linguistic difficulties; which give no linguistic training; which do not illustrate principles; which merely illustrate principles; which work out exactly; which do not work out exactly; which can be solved by definite methods; which cannot be solved by definite methods; which contain trivialities; which refer to large transactions beyond the children's experience—and indeed many other kinds which the critic of the moment happened to dislike. Then in the

[21] *The Psychology of Arithmetic*, pp. 93-94.

name of the great Cocker, what sort of problems *are* we to teach?

It is always easy (and sometimes profitable) to poke fun at problems. Stephen Leacock does so to great effect. He regards them as stories with a plot. The characters of the plot are people named A, B, and C, who are generally engaged on a mysterious task, vaguely described as "a piece of work." Sometimes, however, the job is defined more clearly, and A, B, and C are set to dig trenches, or mow fields of hay, or pump water, or to compete in walking matches. They are creatures of different temperaments and different capacities. "A is a man of great physical strength and phenomenal endurance. He has been known to walk forty-eight hours at a stretch, and to pump ninety-six. His life is arduous and full of peril. A mistake in the working of a sum may keep him digging a fortnight without sleep. A repeating decimal in the answer might kill him." "B is a quiet, easygoing fellow," and C is "an undersized, frail man, with a plaintive face." Are we not constantly being told that A can do as much work in 3 hours as B can do in 5, or C in 10? Mr. Leacock relates the sad story of C's death. His frail constitution broke down under the strain put upon it by the arithmetic books. He took to his bed and called in a doctor who had been "reduced to his lowest terms." "C's life might even then have been saved, but they made a mistake about the medicine. It stood at the head of the bed on a bracket, and the nurse accidentally removed it from the bracket without changing the sign." [22]

[22] *Literary Lapses,* pp. 237-245.

These "time and work" sums are of little practical value. They assume a uniformity of output which is alien to human nature. No man can perform six times as much work in six consecutive hours as he can in the first hour. Nor can two men working together be trusted to do twice as much work in a given time as they can do separately. This type of problem, however, illustrates the fact that although we cannot add and subtract times when the work is fixed, we can add and subtract quantities of work when the time is fixed.

To a similar class belong cistern problems. Here we generally have a cistern fed by two taps and emptied by a third. Some idiot comes along and turns on the three taps at once; and we are asked to measure the magnitude of his folly. "If a cistern can be filled by one tap in 6 minutes, filled by another in 10 minutes, and emptied by a third in 8 minutes, how long will it take to fill the cistern if the three taps are kept running at the same time?"

The excuse for this outrage on common sense is that it seems the sole sort of problem that affords an opportunity of subtracting as well as adding quantities of work. We cannot put one man to mow a field and another man to un-mow it. Something might possibly be done with the digging of ditches. While a navvy or two might be set to shovel out the dirt, others might be set to shovel it back again, and the problem would be to determine the state of the ditch at the end of a given time. But nobody has yet asked so fantastic a question—not even in an examination paper.

In spite of the critics, most of the older teachers have a secret affection for A, B, and C, and find little solace in seeing their jobs transferred to the algebraic department and done by blacklegs, *a*, *b*, and *c*.

I wish to break a lance with the critics over another point—their attempt to make arithmetic a medium for teaching English. Let us glance at the following extract from a recent report issued by the Board of Education[23]:

"In no Scheme that we have seen is there any indication of the possibility of using the Arithmetic lesson as a means of training in English. Yet the exercise involved in stating accurately the nature of a problem and the method to be employed in its solution is not only a necessary preliminary to mathematical calculation, but a most valuable means of securing clear thought and lucidity of expression."

Here is a case of enthusiasm run wild. Can anybody, with an ounce of psychological discernment, seriously hold that we must of necessity be able to put clearly and accurately into words the nature of a problem and the steps to be employed in solving it before we can begin to calculate? I am myself as ready to believe that a boy cannot digest his dinner till he can give a clear and accurate account of the process of digestion and of the chemical changes that take place in the conversion of food into blood; or that he cannot speak before he can understand the laws that regulate the production of articulate sounds, and can appreciate and formulate the rules of grammar. As a matter of fact, mathematical

[23] *General Report on the Teaching of English in London Elementary Schools*, sec. 21.

calculation of a simple sort is well within the capacity of a child of seven, but to render an account of the mathematical processes he employs is a task which requires far more mature powers, and is in any case a subsequent analysis rather than a necessary preliminary.

This excessive insistence on words is but froth and foam from that strong current of enthusiasm for the mother-tongue which found its amplest expression in the Departmental Report on the *Teaching of English in England.* I yield to none in my admiration of that document, and I accept, as the new *Suggestions for Teachers* accepts, Mr. George Sampson's dictum that every teacher *in* English is a teacher *of* English, but I am not prepared to admit that English is the be-all and end-all of school instruction. Nor is it the only language in the school. To say nothing of foreign languages, there is another language more universal than any of them—more precise, compact, and unequivocal than any other tongue ever spoken or written—the language of mathematics. And it is this language, not the English language, that we should try during the arithmetic lesson to get our pupils to speak and write with understanding and with facility. As the good teacher of French is economical of his English and lavish of his French, so does the good teacher of arithmetic aim, above all things, at getting the maximum of arithmetic done in the arithmetic time. With intrusions upon that time he has little patience.

Note, however, what the *Handbook of Suggestions for Teachers* (p. 188) says:

"The children should be encouraged to write one or more suitable words opposite each line of these simple sums, e.g. if the question demands the price of two clocks at seven pounds each, the words 'cost of 1 clock' should appear opposite the multiplicand, and 'cost of 2 clocks' opposite the answer. If the signs +, −, ×, ÷, and = have been taught, as some teachers prefer, then the children may write:

Cost of 1 clock = £7
Cost of 2 clocks = £7 × 2 = £14."

I see no reason to object to a child in the infant school indulging in this luxury of expansion; for there the study of numbers should proceed at a leisurely pace, and there it does not much matter if things are mixed up somewhat, and he is taught much handwriting and much composition and a little arithmetic during a lesson which is called "number." But when he reaches the senior school he has to tackle arithmetic in grim earnest. If a child is so young as to be profitably engaged in applying the "twice" table, it will take him no small amount of time to write "cost of 2 clocks," and, whatever his age, it will take him at least ten times as long as it will to write the figures 14. Besides, the extra verbiage is neither number nor reason. If a child can answer the question at all, he knows quite well that the £14 means the price of 2 clocks, and not the price of 5 motor-cars, nor of 12 chickens. In the writing down there is more manual labour than mental exercise. And it takes up the time of a dozen oral questions.

Teachers, to do them justice, have realised this, and

do not nowadays yield to this particular "suggestion." They know better. They have tried it in the past and know that it has led to the very state of affairs about which the Board's inspectors and others have so justly complained—an unsteady grip of the very rudiments of arithmetic. They tried it in the past just as they tried that other piece of advice so freely and so confidently given a generation ago: the child should always answer in complete sentences. It was not only found to be unnatural: it was also found to be a waster of time and of temper.

A larger and more acceptable aftermath of the English Report appears as an Appendix to another of the Board's publications—*Some Suggestions for the Teaching of English in Secondary Schools in England.* This Appendix shows how the study of mathematics may contribute to a training in English. With the writer of this persuasive article one is far more inclined to agree, partly because the English he refers to is mainly oral, and partly because the pupils with whom he is concerned have already acquired a fair measure of facility in the use of mathematical language proper.

The view that English may be taught through arithmetic has this much of truth in it. The science of arithmetic, like any other science, has a jargon of its own, and it is a jargon that the pupil must learn. He must learn that the word "product" has one meaning in industry and another in arithmetic, and he must learn that the symbol × has not quite the same meaning in a sum as in a letter. It is, in fact, an essential part of a child's training in arithmetic to learn the technical meaning of such terms as Average, Percentage, Ratio, Rate percent, Interest, Significant figures, and so on. But whether this is learning English through arithmetic or arithmetic through English, I am not quite sure.

What are we to think of the long and wordy problems that appear in the examination papers for a First School Certificate? Look at this specimen, which is fairly typical:

"A and B combine to buy a roll of cloth containing 60 yards at 10*s.* per yard; A is to have two-fifths of the cloth, and B three-fifths. When the cloth has arrived, B lets C have a quarter of his share, and A lets C have a third of his share, each at cost price. Three-eighths of what A now has left is transferred by him to B at a profit of 20 percent. How many yards has each of the three for his own use, and what is the net cost to each of his final share?"

I consider this sort of question altogether bad—bad as a test and bad as training. It is bad as a test because it tries to test too much at once. It tests general intelligence, it tests English, and it tests arithmetic; and it tests neither of them well. It requires for its solution the same sort of capacity as is needed for an "instructions" test such as this: "If an elephant is larger than a kangaroo, put a cross in the bottom left-hand corner of this paper, unless four is larger than three times the third of three-and-a-third, in which case underline the word that comes before the word that comes after the word 'cross' in this sentence." This, of course, is only a roundabout way of asking you to underline the word "cross"; but to arrive at that conclusion requires an average intelligence and a careful and patient reading of the sentence. The same is true of the arithmetical problem quoted above. The candidate must patiently and doggedly disentangle the meaning of the words. He will then find that the arithmetic involved, in spite of its apparent complexity, is quite simple and commonplace. It is, however, not a good arithmetic test, because it involves a number of more or less distinct operations, the later of which depend for their data on the results of the earlier. An error in the first will bring disaster on all that follow. An examiner with a conscience will have an uncomfortable time in marking that sum.

As for the view that hard problems afford good exercise for the mind—that they sharpen a child's wits—I have already shown how untenable that position is. The real facts are well put by Mr. George Sampson, who, after quoting examples of problems much simpler than mine, goes on to say: "I ask teachers to examine their experience honestly and say whether they have detected the least increase in the general or specific intelligence of any child resulting from a course of such sums? Is it not simply the case that the boy who is 'good' at that kind of work continues good, while the boy who is 'fair' continues fair—that, in fact, there has been, and will be, no accession to the intelligence of either? Elementary schoolmasters who believe that arithmetic will make children sharp are on just the same plane as public schoolmasters who believe that Latin will make children write English. Such sums as those quoted above are neither tonic nor nutrient." [24]

I have not yet answered the question: What sort of problems should we teach? The reason is that I don't know. I do, however, know what sort of problems we actually will teach: those that are set in the public examinations of the day. And these are probably just as good as any others.

[24] *English for the English*, p. 98.

CHAPTER XXII

RULE OF THREE

Adams: But, Sir, how can you do this in three years?

Johnson: Sir, I have no doubt I can do it in three years.

Adams: But the French Academy, which consists of forty members, took forty years to compile their Dictionary.

Johnson: Sir, thus it is. This is the proportion. Let me see; forty times forty is sixteen hundred. As three is to sixteen hundred, so is the proportion of an Englishman to a Frenchman.

Boswell: *Life of Johnson.*

THE sense of the above quotation is clear, though the ratio is wrong. The inversion was probably due to a lapse of memory on Boswell's part; for Johnson himself was a good arithmetician. Mrs. Thrale tells us that he used to practise calculation "when he felt his fancy disordered," and Boswell records, with some surprise, that his hero once presented a Highland lassie with a copy of *Cocker's Arithmetick.*

Rarely is rule of three taught nowadays as Cocker taught it. The old method of proportion has gone out of use: it has been supplanted by the method of unity. Upon the value of this method two useful pieces of research have been carried out. The first is by Mr. W. H. Winch,[25] who demonstrates the soundness of

[25] *The Journal of Experimental Pedagogy,* vol. ii, Nos. 2-6 (1913-14).

the method, and shows that if small numbers are used, proportion may profitably be taught at an earlier age than is now customary. The second is by Miss Jeannie B. Thomson,[26] who contends that although the unitary method is good, the fractional method is better, and that therefore the fractional method should be the final goal. With that conclusion I agree. Let me take a simple example and work it by both methods in order to bring out clearly the difference between them.

Example I.—If 5 lb. of cheese cost 4*s*. 2*d*., how much will 15 lb. cost?

(a) Method of Unity:

5 lb. cost   4*s*. 2*d*.

1 lb. costs $\dfrac{4s.\ 2d.}{5}$

15 lb. costs $\dfrac{4s.\ 2d.}{5} \times 15$

(b) Fractional Method:

5 lb. cost   4*s*. 2*d*.

15 lb. cost   4*s*. 2*d*. $\times \dfrac{15\ \text{lb.}}{5\ \text{1b.}}$

It may be contended that the second method is the same as the first with the middle step left out, for the final statement in both cases is $\dfrac{4s.\ 2d. \times 15}{5}$.

It is true that the final results are identical (otherwise both methods could not be valid), yet the

---

[26] *The Art of Teaching Arithmetic,* chap. xii.

mental processes involved are widely different. The notion of the price of one thing is present in the first and absent in the second, and the notion of ratio—of a direct comparison between two magnitudes—is present in the second and absent in the first.

It seems more natural to employ the unitary method in some examples and the fractional method in others. Note these two examples:

Example II.—If 3 tons of coal cost £6, what will 17 tons cost?

Example III.—If 3 tons of coal cost £5 18*s.* 3*d.*, what will 12 tons cost?

II entices us to say: One ton costs £2, therefore 17 tons cost £34; and III just as strongly impels us to say: Four times the quantity will cost four times the price, therefore the answer is £23 13*s.* It is the second style of question that brings out the notion of ratio.

Let us now put forward the fractional method as the standard rule for working proportion. It is very simple, and I have good grounds for believing that it is very effective. There are two steps:

1. Put down the term that is like the answer.

2. Multiply by a fraction formed of the other two terms. The second step is ambiguous, as from two terms, a and b, we can form two distinct fractions, $\frac{a}{b}$ and $\frac{b}{a}$. Which of the two are we to use? The one which accords with the reply to this question: Is the answer to be greater or less? Now note how simply it works:

Example IV.—If the wages of 12 men for a week amount to £32, what will the wages of 7 men amount to?

1st Step: Put down £32, since the answer is money.

2nd Step: Multiply it by $\dfrac{7 \text{ men}}{12 \text{ men}}$ or by $\dfrac{12 \text{ men}}{7 \text{ men}}$.

This is the same as multiplying by $\frac{7}{12}$ or by $\frac{12}{7}$. As the answer has to be less than £32, I choose the $\frac{7}{12}$.

The Working:

$$£32 \times \frac{7}{12} = \frac{£32 \times 7}{12} = \frac{£8 \times 7}{3} = \frac{£56}{3} = £18 \; 13s. \; 4d.$$

Example V.—If a watchmaker buys 23 watches for £40 16s. 6d., how many of the same kind can he buy for £122 9s. 6d.

1st Step: 23 watches.

2nd Step: $23 \text{ watches} \times \dfrac{£40 \; 16s. \; 6d.}{£122 \; 9s. \; 6d.}$ or $\dfrac{£122 \; 9s. \; 6d.}{£40 \; 16s. \; 6d.}$

As the answer has to be more than 23, I choose the latter fraction.

The Working:

$$23 \times \frac{£122 \; 9s. \; 6d.}{£40 \; 16s. \; 6d.} = 23 \times \frac{4899 \text{ sixpences}}{1633 \text{ sixpences}} = 23 \times 3 = 69.$$

Example VI.—If 7 men mow a field in 5 days, in how many days will 11 men mow it if they work at the same rate?

1st Step: 5 days.

2nd Step: $5 \text{ days} \times \dfrac{7 \text{ men}}{11 \text{ men}}$ or $\dfrac{11 \text{ men}}{7 \text{ men}}$.

As the time will be less than 5 days, I choose the former fraction.

The Working:

$$5 \text{ days} \times \frac{7}{11} = \frac{5 \times 7}{11} \text{ days} = \frac{35}{11} \text{ days} = 3\frac{2}{11} \text{ days}.$$

We should aim at getting our pupils ultimately to adopt the fractional method as the standard rule-of-thumb method of working proportion, and our introductory exercises should foster that aim. We should get the children on the king's highway as soon as possible. In furtherance of this purpose we should familiarise them with the notion of ratio; we should give numerous examples, both written and oral, in which the idea of ratio is prominent; and we should, when the method of unity is used, discourage working out until the final statement is set forth in fractional form.

We should develop the notion of ratio. A ratio is a comparison—a special sort of comparison. I can compare two magnitudes by finding out how much larger one is than the other, or by finding out how many times as large one is as the other. The latter is the ratio way. The distinction may be illustrated by the problem of the father who is 35 years old and has a son 5 years old. He is 7 times as old as his son. In 5 years' time he will be 40 and his son 10; that is, he will be 4 times as old as his son. In another 5 years' time he will be 3 times as old as his son, and 15 years afterwards he will be only twice as old as his son. The question is: How long will these two have to go on living together before they get the same age? As we learnt, when dealing with subtraction, equal additions do not affect the difference (there will always be 30 years difference between this father and his son), but they do affect the ratio.

Since a ratio is a comparison the two terms of a ratio must be the same in kind. Turn back to Example I above. By this method of unity we arrive at the fraction $\frac{4s.\ 2d.}{5}$ This is not a ratio. There is no means by which we can compare 50 pence with the abstract number 5. The two belong to entirely different categories. We can simplify the fraction and say it is equal to 10*d.*, but the 10*d.* implies no comparison of any sort. In Example II, on the other hand, we have a true ratio, $\frac{15\ lb.}{5\ lb.}$ Its value is 3. The 3 implies that the quantity in the numerator is three times as great as the quantity in the denominator. A ratio is always abstract. It is unaffected by the quality of the terms. $\frac{3\ men}{5\ men}, \frac{£3}{£5}, \frac{3\ tons}{5\ tons}, \frac{3\ hrs.}{5\ hrs.}, \frac{3\ yds.}{5\ yds.}, \frac{3\ half\text{-}inches}{5\ half\text{-}inches}$ are all precisely equal. They are all equal to ⅗. Sir Oliver Lodge explains it in this way: $\frac{3\ men}{5\ men} = \frac{3 \times men}{5 \times men}.$ The men cancel out, leaving the ratio in the simpler form ⅗.

Note the equivalence of the three signs ÷, −, and : The first is the complete sign for division, the second has the dots missing, and the third has the line missing; but they all mean the same. So $3 \div 5 = \frac{3}{5} = 3 : 5$. The vinculum of a fraction is a mutilated division sign; and so is the symbol for ratio. In dealing with the equivalence of fractions (see Chapter XVII), we have at the same time been dealing with the equality of ratios, and the equality of ratios is another name for proportion. ½ = ¼ may be written 1 : 2 :: 2 : 4, and read "as one is to two, so is

two to four." Just as : is a mutilated division sign, so is
: : a mutilated equality sign. It is = with the middles of
the two lines left out and only the four ends remaining.[27]

It is now possible to show that the fractional method,
and the ratio method, and the equivalence-of-fractions
method are at root the same. Example I above may be
worked thus: $\dfrac{5 \text{ lb}}{15 \text{ lb.}} = \dfrac{4s.\ 2d.}{x} \therefore \dfrac{1}{3} = \dfrac{50d.}{x} \therefore x = 50d. \times 3$
$= 150d. = 12s.\ 6d.$

The other examples admit of similar treatment.
Although the form, or the formula, of proportion has
gone out of date, the principle of proportion remains
for all time. a : b : : c : d may become obsolete, but
not the meaning of the statement, nor the laws which
it embodies. Indeed, the principle of proportion is
postulated in the very method that is supposed by
some to supplant it. For the method of unity takes it
for granted that certain magnitudes vary together step
by step; that as, for instance, the quantity of goods
increases, the price increases, as the quantity diminishes,
the price diminishes too—and in the same proportion.
It assumes that as the number of men engaged on a job
goes up or down, so does the amount of time taken go
down or up—and in the same proportion. And this
very question of concomitant variation or proportional
change, upon which the validity of our solution depends,
cannot always be taken for granted. It cannot be taken
for granted in actual life as frequently as it is in textbooks.
We cannot assume in business life that the man who
charges a penny for a lead-pencil will charge twelve

[27] If this is not what : : means, it is what it ought to mean.

shillings for a gross; nor can we assume that if a 1-lb. jar of honey costs 10½*d*., a 3-lb. jar will cost 2*s*. 7½*d*. It is quite legitimate, for computation purposes, to assume that two bricklayers working together will lay twice as many bricks as one of them working by himself, or that a train that travels fifty miles in the first hour will travel exactly a hundred in the first two hours, or even that if 10 men working 8 hours a day and 6 days in the week can build a house in 6 months, 15,000 men could build it in 5 minutes. But useful as these assumptions are for the purpose of rough calculation, life itsell plays ducks and drakes with them. It puts in all sorts of provisos and accidents, which prevent our conclusions from being absolutely true: they are only conditionally true—only true if certain things take place, which never, except by a miracle actually do take place.

Besides these legitimate assumptions of proportionality there are others which are illegitimate and to which the young beginner may easily be betrayed. For example: If by selling a knife for half a crown I gain 6*d*., what must I sell it for to gain 1*s*.? If we assume the selling price to be proportional to the gain, we get the absurd answer, 5*s*.

Children may be made judiciously wary by occasionally mixing a few spurious proportion sums with genuine ones. Here are a few, culled from various sources:

1. If it takes 3 minutes to boil an egg, how long will it take to boil 10 eggs?

2. If the diameter of a half-penny is 1 inch what is the diameter of a threepenny bit?

3. If a barking dog keeps 2 men awake all night, how many barking dogs will keep 11 men awake for 3 nights?

4. If it takes 2½ secs. for a man to fall down a precipice, how long would it take 20 men to fall down the same precipice?

5. If John is 4 feet high when he is 10 years old, how tall will he be when he is 40?

6. If Henry the Eighth had six wives, how many wives had Henry the Fourth?

These are flagrantly absurd, and can scarcely deceive the meanest intelligence. Others are more subtle. For example:

7. If the wire joining 10 telegraph poles equally spaced is 585 yds. long, what length of wire joins 5 of them?

8. If a stone falls 144 ft. in 3 seconds, how far will it fall in 6 seconds?

9. If the area of a circle of 3 inches diameter is 7.0686 square inches, what is the area of a circle whose diameter is 6 inches?

In the 7th question the ratio is not 10 to 5, but 9 to 4. To No. 8 the answer is not 288 feet, but 576 feet. The distance does not vary with the time, but with the square of the time. The same law applies to No. 9. When the diameter is doubled, the area is not doubled, but quadrupled.

Thus the study of the equivalence of fractions leads to the study of proportion, and the study of proportion to the study of function. One quantity is a "function" of another quantity when it changes as that other quantity changes—not arbitrarily, but according to a fixed law. It need not be a simple law, as in direct or indirect proportion, but it may be expressed by a mathematical formula and permits of exact calculation. The formula $S = \frac{1}{2}$ ft.^{2}, by which No. 8 can be solved is a case in point.

The fact that the areas of similar figures vary as the squares, and the volume as the cubes of their linear dimensions, is a piece of knowledge not very widely diffused. At the beginning of the Great War the papers were prone to give pictorial representations of the relative sizes of the belligerent armies. A small soldier stood for the army of one country, and a large soldier three times as tall for the army of another. It was never quite clear whether the drawings were meant to convey the information that the second army was three times as large as the first, or nine times, or twenty-seven times. If the areas of the surfaces covered by the drawings were meant to be compared, the correct interpretation was nine times; if the solid bulk of the men, twenty-seven times.

Here is an interesting breakfast-table problem. Try it on a youth who prides himself on his mathematics:

If I have two apples of precisely the same shape and the same substance, but the diameter of one is 1 inch and the corresponding diameter of the other 3 inches, how many times is the larger apple as heavy as the smaller?

CHAPTER XXIII

INCOMMENSURABLES[28]

> Nor cut thou less nor more
> But just a pound of flesh: if thou cut'st more
> Or less than a just pound, be it but so much
> As makes it light or heavy in the substance,
> Or the division of the twentieth part
> Of one poor scruple, nay, if the scale do turn
> But in the estimation of a hair,
> Thou diest and all thy goods are confiscate.
>
> SHAKESPEARE: *Merchant of Venice.*

IF I carefully count the apples in a basket, I can feel quite sure about the accuracy of the result. I know I am right; not nearly right, but quite right. If, however, I weigh one of the apples, I have no such feeling of confidence. I am not sure that the weights are true or that the scales are accurately adjusted. If I use a more delicate balance or a finer gradation of weights, I get a different result. I have weighed as precisely as I can, but I am not prepared to swear to the absolute accuracy of my estimate. Weighing admits of degrees of precision:

---

[28] Those who are interested in this topic are advised to read chap. xx of Sir Oliver Lodge's *Easy Mathematics,* and chaps. lxxi and lxxii of Professor Nunn's *Algebra.*

counting does not. Counting gives us a number which is either quite right or quite wrong; measuring gives us a number which may be right as far as it goes, but it doesn't go the whole way. It tells us nothing but the truth, but it doesn't tell us the whole truth.

The same remarks apply to the palings of a fence. I can count them, and count them correctly, but if I measure one of them, I get a result which is only approximate. If I measure it twice and get the same result each time, it points to either looseness in my observation or coarseness in the instrument I use. Closer observation or a finer instrument would give measurements which consistently vary. The most I can say of a given measurement is that it is as accurate as the means at my disposal will permit.

The difficulty of exact measurement comes clearly to light when we consider the ancient problem of doubling the area of a square. Let ABCD represent a square, each side of which is 1 inch. If I wish to construct a square whose area is twice as large, that is, two square inches, how long must its side be? The

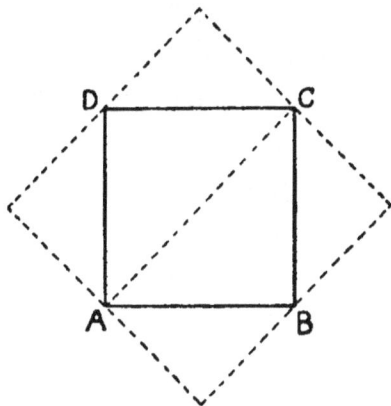

answer, of course, is, "As long as the diagonal of the square." But how long is the diagonal? How long is AC? If the reader will try to ascertain by measuring, he will find that the length is somewhere between 1.4 and 1.5 inches. With an ordinary ruler he can scarcely get closer than that. We know by reasoning that the real length is $\sqrt{2}$ inches. But 2 has no square root: we can find no number, integral or fractional, which when multiplied by itself will make 2. Indeed, it can be proved that there is no such number. If we extract the square root in the ordinary way, we find it to be 1.41421 ... The calculation may go on for ever. It will neither terminate nor produce a recurring decimal. Let us take the first two figures. 1.4 squared equals 1.96, and this differs from 2 by .04. Taking the first three figures and squaring their value, we get 1.9881, a closer approximation to 2. Four figures give 1.999396, five figures 1.99996164, six figures 1.9999999241. It will thus be seen that although we can never find the exact square root of 2, we can find a series of numbers which get nearer and nearer to it, and are thus able to express it with an increasing degree of accuracy. Indeed, they express it as closely as the number of figures will permit. The nearest approach to $\sqrt{2}$ that can be expressed by two figures is 1.4, the nearest that can be expressed by three figures is 1.41, by four figures 1.414, and so on.[29] We express these facts by saying that $\sqrt{2}$ is an irrational number, and that the diagonal of a square is incommensurable with the side. This is another way of saying that if we take the side of a square as a unit, we cannot express the length of the

[29] A pupil of Augustus De Morgan's carried the value of $\sqrt{2}$ to 110 decimal places. See *A Budget of Paradoxes,* ii, 68.

diagonal exactly in terms of that unit or of fractions of that unit. Conversely, if we take the diagonal as the unit, it is the side that becomes incommensurable.

In the same sense the circumference of a circle is incommensurable with the diameter, or, which is the same thing, the diameter is incommensurable with the circumference. If the diameter is 1, the circumference is 3.14159 . . . If the circumference is 1, the diameter is .3183 . . .

When we measure real things by applying such standard units as feet, yards, metres, or indeed any other arbitrary unit we care to select, we find the same sort of incommensurability. Our estimates are always approximate, never exact. Sir Oliver Lodge puts it thus:

"Incommensurable quantities are therefore by far the commonest, infinitely more common, in fact, as we shall find, than the others; 'the others' being the whole numbers and terminable fractions to which attention in arithmetic is specially directed, which stand out, therefore, like islands in the midst of an incommensurable sea; or, more accurately, like lines in the midst of a continuous spectrum."

It is by some such considerations as these that we can bring home to the pupil the need for means of expressing degrees of approximation, and for means of dealing with approximate numbers. He can be made to see the point of talking about significant figures. Consider the following measurements: 307,000 metres, 3070 metres, 30.7 metres, 3.07 metres, .307 metres, and .0307 metres. They are all 307 something—307 kilometres, or decametres, or what not. The figures are the same, the units are different. All the measurements are expressed in those significant figures. If the real value were 306 and not 307, the inaccuracy, or the "error," would be relatively the same for all the measurements. "Three significant figures" is, in fact, a phrase indicating a certain degree of accuracy. "Four significant figures" would indicate a still higher degree of accuracy, and so on. When we require an answer to three places of decimals, we are asking for a certain degree of absolute accuracy; when we require it to three significant figures, we are asking for a certain degree of relative accuracy. And it is relative accuracy that really matters. In buying a ton of coal, an ounce or two one way or the other is not of much consequence. In buying a pound of tea or of tobacco, an ounce is a real consideration. An inch added to a man's nose is a very different thing from an inch added to the Atlantic cable.

We have seen which are the significant figures; but which are the non-significant figures? Clearly those which are mere place-keepers—those which do not serve as indicators of exact magnitude, but simply give correct value to the other figures. There is only one figure that satisfies this condition, and that is zero. Nought is sometimes, then, a non-significant figure; but not always. It is significant if it definitely asserts that there is nothing there—if it guarantees the value of a particular denomination to be zero. I vaguely recollect having read somewhere that Charles Dickens died worth £90,000. None of these noughts are significant: they merely label the 9. For what I mean by the statement is that in the roundest of round numbers—a number which can be expressed by one significant figure—Charles Dickens bequeathed that sum of money. I mean that the remembered sum is nearer the mark than either £80,000 or £100,000. At this point I look up Forster's *Life of Charles Dickens* and find that "the real and personal estate amounted, as nearly as may be calculated, to £93,000." Forster estimated the amount to two significant figures, I recollected that estimate to one. Neither Forster's noughts nor mine signify anything: they do not deny the possibility that the amount of the estate was £92,864.

There is only one contingency when a nought is always significant, and that is when it is sandwiched between two digits; and only one when it is always non-significant, and that is when it is a mere place-keeping prefix. The noughts are significant in 508,007 and in 4.06; they are nonsignificant in 0.34 and in 0.008.

When the last figure of a decimal is nought, there is a strong presumption that the nought is significant. If a book is priced at 3.50 dollars, the 50 means precisely 50 cents., not 51, nor 53, nor 48. If a line has been measured and estimated at 0.170 metres, it suggests measurement to the nearest millimetre. If it had been measured to the nearest centimetre, the result would have been given as 0.17 metres.

Finally, we have the case in which an integer ends in noughts. We cannot know, unless we are told, whether the noughts are significant or not. We can sometimes guess, though. If I am told that a certain house has been bought for £1500, I presume that the noughts are significant. If I am told that an author made £1500 by his books last year, I suspect the noughts to be non-significant.

It is in laboratory measurement, however, that significant figures become supremely important. Sir Oliver Lodge assures us that the result of such measurement is *always* an incommensurable number. The value of a piece of scientific research often depends on increasing the number of figures upon which reliance can be placed. Not that we can ever go very far by direct measurement. To quote Sir Oliver Lodge once more[30]: "A few exceptionally skilled experimenters with a genius for the work, devoting a year to a research, might attain 5-figure accuracy, but such accuracy as this is generally limited to the astronomical observatory, where the measurements are fairly simple and the theory of the errors to which instruments are necessarily liable has been studied for centuries. In taking the mean of a number of astronomical observations, even 6-figure accuracy is attainable, but beyond this it is extremely difficult to go."

In the attempts to find the value of π, direct measurement proved of little use. By this simple means the value cannot be estimated beyond a few places of decimals. Archimedes had to resort to a form of geometrical reasoning, which depends upon the obvious fact that the circumference of a circle must be less than the perimeter of a circumscribed polygon, and greater than the perimeter of an inscribed polygon, however many sides these polygons may have. By this

[30] *Easy Mathematics,* p. 195.

method, Ludolf van Ceulen (1539-1610) calculated π to 35 places, and had the figures carved on his tombstone. Finally, algebraic series were employed, and it became possible for Shanks in 1873 to publish the value of π correct to 707 places of decimals.

There are evidently two kinds of incommensurables—those which belong to pure abstract arithmetic and those which belong to applied arithmetic. We not only know that $\sqrt{2}$, $\sqrt[3]{2}$ and log 2 are incommensurable, we know that they *must* be incommensurable. We cannot, however, feel so confident about the result of measurement. In spite of the dictum quoted above, we feel that the quantity measured *might* be commensurable with the unit of measurement. The odds are overwhelmingly against it: but it is not impossible.

These facts bring out the difference between the discontinuous things, which we can count, and the continuous things, which we have to measure. They bring out the vital difference between the concrete arithmetic with which the learner begins and the concrete arithmetic with which he ends. He begins in the kindergarten with counting sticks and beads, he ends in the laboratory with weighing salts or measuring the temperature of liquids. The aim of the first is to give clear notions of numbers and of the operations that may be performed on them, the aim of the second is to apply those notions to the material world in which we live. Between these two comes the abstract arithmetic of the textbook. In this abstract realm we deal with hard uncompromising concepts and hard uncompromising facts. Here everything is neat and trim and perfect.

There is no vague merging of one thing into another, no gradual fading away into obscurity. That 2 and 2 make 4 admits of no doubt or cavil; that ¾ of 12 is 9 no one who knows the meanings of the symbols is ever likely to dispute; that a quarter of a foot is a twelfth of a yard is a conclusion that is reached by irrefutable logic from the defined relationship between a foot and a yard. The learner asserts with confidence that if one pound of tea costs l*s.* 10*d.*, three pounds of the same tea will cost 5*s.* 6*d.*, quite undisturbed by any views regarding the possibility of weighing out exactly one pound of tea, or of weighing out another quantity precisely three times as large. That consideration is as little relevant as is the question whether the purchaser can afford to pay for the tea. The "if," at any rate, saves the situation. In this Platonic realm of numerical ideas the assumption of perfect relationships is not only legitimate, but necessary. We cannot calculate without them. We can make what discounts or reservations we like after the calculation is over: we must not intrude them upon the calculation itself.

The theory of irrational numbers and incommensurables belongs to advanced arithmetic, and the question arises: How far is it to throw its shadow down the course of study in our schools? We see its influence extending lower and lower. It falls heavily on the secondary school: it is not unfelt in the primary school. My own view is that it extends too far. The pupil of eleven, or indeed before thirteen, should not have his arithmetical notions clouded by irrationals, nor should he have his mind muddled by methods whose main justification is that they are well suited for incommensurables. How can we account for the growing tendency to exalt decimal fractions (with which I sympathise) and to decry simple vulgar fractions (with which I do not sympathise)? Presumably because decimal fractions lend themselves to approximate methods, while vulgar fractions do not. The plea that a decimal fraction, deriving as it does from our ordinary notation for integers, is easier to understand than a vulgar fraction finds support neither in history nor in psychology. Vulgar fractions are as old as arithmetic itself. The most ancient treatise on mathematics which we possess (that of Ahmes the Egyptian on *Directions for obtaining the Knowledge of all Dark Things*) begins straight off with vulgar fractions. Of the dark things with which he dealt, he evidently did not regard vulgar fractions as the darkest. Be that as it may, it is clear that vulgar fractions were known three or four thousand years before decimal fractions were even thought of. And the child, recapitulating in a broad way the history of the race, apprehends vulgar fractions before he can grasp the significance of decimal fractions. With the aid of apparatus and concrete material the young child in the infant school can deal quite intelligently with halves and quarters and thirds, but it is doubtful whether at this tender age he can see the full meaning of our decimal notation. It is true that he may be brought to understand the value of each digit in such a number as 37. But so could the Greeks and the Romans. Decimal notation had been in use for thousands of years before the principle of position was adopted. And just as the

THE ESSENTIALS OF ARITHMETIC

human race found it more difficult to evolve a theory of local value than a theory of fractions, so does the child find it more difficult to grasp the notion of place value than to understand the nature of a simple fraction. That is one reason why I hold the view that the multiplication of decimals should be approached from the fractional side rather than from the notational side. It should be approached from the notational side too, but when the learner's mind is a little more mature.

An allied tendency, flowing from the same source, is the exaction of a preliminary estimate. A rough approximation, useful as it is in long and complicated calculations, is not an advantage in each and every example. And it is altogether a teacher's notion, not a child's notion. Neither the bright child nor the dull child takes kindly to it. He evades it if he can. Apparently it does not seem to him to be worth while. Every examiner knows that in spite of repeated counsels to make a rough preliminary estimate, the young candidate goes his own way and leaves it out. He is, in fact, in that middle stage to which I have referred above. The notion of degrees of accuracy is alien to his young mind. A sum is either right or wrong, and that's an end of it. And if it is wrong, it might just as well be wrong by a pound as a penny. Hence the preliminary guess does not appeal to him as it does to the adult mind. To him it means getting the sum wrong first and getting it right afterwards. It means telling a lie as a preliminary to telling the truth. This, at any rate, seems to me to be the clue to the child's undoubted reluctance to give two answers to the same sum.

I am not arguing against a preliminary guess; I am simply trying to explain the young pupil's resistance, and counselling the teacher to patience.

The reluctance of the modern teacher of mathematics to retain the traditional rule for the multiplication of decimals (a topic which I have almost worn threadbare) seems to be rooted in a conviction that a decimal fraction is an expression which is of large and definite value at one end, but tails off into nothingness at the other end, and that the only way to deal adequately with it is to grip it firmly at the large end, and have as little as possible to do with it at the small end. There is in this view much common sense, especially if the decimal is the result of measurement (and therefore approximate), and if contracted methods are to be encouraged. But it is not a view to be forced upon the beginner. To him one end is just as definite as the other. And for purposes of calculation he is justified in his view. Even if the number is really incommensurable, and, because it must stop somewhere, stops at the third place of decimals, it must be treated as though the whole value were completely expressed. He should begin by accepting the maxim, Take care of the pence and the pounds will take care of themselves. Later on he will see the folly of being penny wise and pound foolish.

Bearing these reservations in mind, the teacher is well advised to exhort his pupils to make preliminary guesses whenever they are helpful; and they are helpful wherever there is danger of getting lost among the details. The little things are more apt to bewilder than the big things. A prosperous draper who had for many years taken into his employment boys and girls straight from school told me that his favourite test to applicants was this: Find the cost of 7¾ yards at 7¾*d*. a yard. They generally took a long time over it, and often gave an absurd answer. It was the farthings that puzzled them, not the pence. If they had simply increased one quantity by ¼ and diminished the other by the same amount, they would have got 7½*d*. × 8, that is, 5*s*., straight off. It would also be the correct answer, to the nearest farthing.

It is in dealing with decimal fractions where the figures are many but the magnitudes are small that rough estimates are specially useful. Here the small things are peculiarly liable to be confused with the large things; for the small things do not look small. It is a good plan to practise pupils in making diagrams to represent the relative magnitudes of the figures in a decimal, as, for example, in 2.679 (see diagram below). In course of time they will come to picture such an expression as a sum of values that diminish with extraordinary rapidity, and they will cease to be impressed by decimal digits remote from the decimal point.

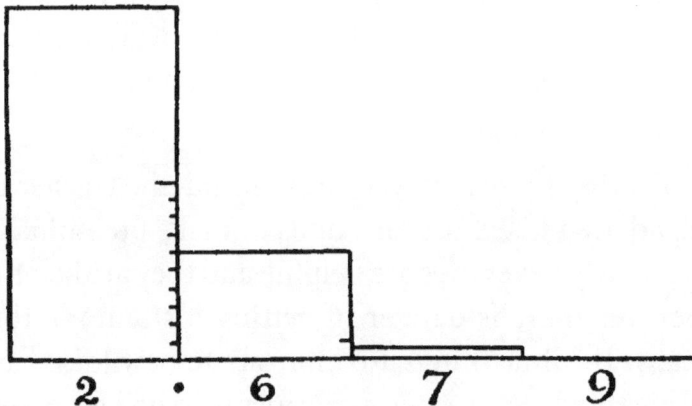

In making a rough estimate, let it always be a matter of ingenuity and intelligence, never a matter of mechanical routine. Work the sum mechanically if you like (it is indeed best to work it mechanically), but check it intelligently. One of my objections to "standard form" methods is that both the preliminary estimate and the complete working are obtained by the same procedure, and one tends to become as mechanical as the other. The two are in collusion and cannot give independent evidence. The greater the difference in the angle of approach and the mode of calculation between the answer and the check, the greater the corroborative value of the check. In the multiplication and division of decimals the most valuable check is afforded by an approximate vulgar fraction. If, for instance, .2685 appears, ¼ may be substituted in making the rough calculation. It is rare that the decimal part of a mixed decimal cannot be roughly represented by ¹⁄₁₀, ⅛, ¼, ⅓,½, ⅔, or ¾. This, however, should not be the fixed rule. For making the preliminary estimate there should be no fixed rule.

The cutting short of a decimal fraction at the tapering end is a valuable device. It gets rid of perplexing quantities, which are nevertheless so small as to be negligible. It is not only irrational numbers and recurrers that we treat in this manner: we deal in the same way with all decimal fractions that run to inconvenient length. Unless the unit is very large, it is rarely necessary to carry a decimal fraction beyond the third place. Where it is cut off, there is, so to speak, a raw edge. It is at that point alone that indefiniteness exists. The total value is regarded as approximate, but the departure from the true value lies solely in the last figure retained. If I measure a line to the nearest millimetre and estimate it at .274 metre, I am certain about the 2 and the 7, but all I can say about the 4 is that it represents the real value better than any other figure. What the estimate asserts is that the true value lies within half a millimetre on either side of .274 metre. In other words, it lies between .2735 metre and .2745 metre.

If we are to accept the dictum that all measurements are incommensurable, how are we to deal with a simple question of this kind: Find the area of a rectangle whose sides are 3.5 centimetres and 2.8 centimetres respectively? Are we to say definitely that the area is (3.5 × 2.8, that is, 9.80) square centimetres? Or are we to argue thus : The area cannot be less than (3.45 × 2.75) square centimetres, and cannot be greater than (3.55 × 2.85) square centimetres; it therefore lies between 9.4875 and 10.1175 square centimetres? The question brings out the difference between pure arithmetic, with which the school child is mainly concerned, and applied arithmetic, which is better suited for advanced study. In pure arithmetic we assume that the lengths of the sides of a rectangle can be given with absolute accuracy. The solution follows the lines of pure logic. All it asserts is that if one side is 3.5 centimetres long and the other side 2.8 centimetres long, then the area is 9.8 square centimetres exactly.

An incommensurable never ends. A book does. This one ends here.

A GUIDE TO
BRITISH MONEY

before decimalisation on February 15, 1971

Symbols

£ for pound comes from the Latin word *libra*, referring to a pair of scales. It was also used to designate a Roman unit of weight.

s for shilling comes from the Latin word *solidus*, a Roman coin.

d for penny comes from the Latin word *denarius*, a Roman coin.

Equivalences

£1 = 240 pennies (240*d*)

£1 = 20 shillings (20*s*)

1*s* = 12 pennies (12*d*)

a crown = 5 shillings (5*s*)

a half-crown = 2 shillings and 6 pennies (2*s* 6*d*)

a florin = 2 shillings (2*s*)

a halfpenny = ½ penny (½*d*)

a farthing = ¼ penny (¼*d*)

www.ingramcontent.com/pod-product-compliance
Lightning Source LLC
Chambersburg PA
CBHW031925190326
41519CB00007B/422